Supporting Innovation Through National R&D Programs:

Emerging Research and Opportunities

Amiram Porath
Independent Researcher, Israel

A volume in the Advances in Business Information Systems and Analytics (ABISA) Book Series

IGI Global
DISSEMINATOR OF KNOWLEDGE

Published in the United States of America by
IGI Global
Business Science Reference (an imprint of IGI Global)
701 E. Chocolate Avenue
Hershey PA, USA 17033
Tel: 717-533-8845
Fax: 717-533-8661
E-mail: cust@igi-global.com
Web site: http://www.igi-global.com

Library of Congress Cataloging-in-Publication Data

Names: Porath, Amiram, author.
Title: Supporting innovation through national R&D programs : emerging research and opportunities / by Amiram Porath.
Description: Hershey : Business Science Reference, [2017]
Identifiers: LCCN 2017019855| ISBN 9781522536529 (hardcover) | ISBN 9781522536536 (ebook)
Subjects: LCSH: Technology and state. | Technology--International cooperation. | Research and development projects.
Classification: LCC HC79.T4 P663 2017 | DDC 338.9/26--dc23 LC record available at https://lccn.loc.gov/2017019855

This book is published in the IGI Global book series Advances in Business Information Systems and Analytics (ABISA) (ISSN: 2327-3275; eISSN: 2327-3283)

British Cataloguing in Publication Data
A Cataloguing in Publication record for this book is available from the British Library.

Advances in Business Information Systems and Analytics (ABISA) Book Series

ISSN:2327-3275
EISSN:2327-3283

MISSION

The successful development and management of information systems and business analytics is crucial to the success of an organization. New technological developments and methods for data analysis have allowed organizations to not only improve their processes and allow for greater productivity, but have also provided businesses with a venue through which to cut costs, plan for the future, and maintain competitive advantage in the information age.

The **Advances in Business Information Systems and Analytics (ABISA) Book Series** aims to present diverse and timely research in the development, deployment, and management of business information systems and business analytics for continued organizational development and improved business value.

COVERAGE

- Data Analytics
- Business Process Management
- Business Intelligence
- Performance Metrics
- Forecasting
- Data Management
- Management Information Systems
- Algorithms
- Decision Support Systems
- Business Decision Making

The Advances in Business Information Systems and Analytics (ABISA) Book Series (ISSN 2327-3275) is published by IGI Global, 701 E. Chocolate Avenue, Hershey, PA 17033-1240, USA, www.igi-global.com. This series is composed of titles available for purchase individually; each title is edited to be contextually exclusive from any other title within the series. For pricing and ordering information please visit http://www.igi-global.com/book-series/advances-business-information-systems-analytics/37155. Postmaster: Send all address changes to above address.

Titles in this Series

For a list of additional titles in this series, please visit:
https://www.igi-global.com/book-series/advances-business-information-systems-analytics/37155

Smart Technology Applications in Business Enironments
Tomayess Issa (Curtin University, Australia) Piet Kommers (University of Twente, The Netherlands) Theodora Issa (Curtin University, Australia) Pedro Isaías (Portuguese Open University, Portugal) and Touma B. Issa (Murdoch University, Australia)
Business Science Reference • ©2017 • 429pp • H/C (ISBN: 9781522524922) • US $210.00

Advanced Fashion Technology and Operations Management
Alessandra Vecchi (London College of Fashion, University of London Arts, UK)
Business Science Reference • ©2017 • 443pp • H/C (ISBN: 9781522518655) • US $200.00

Maximizing Business Performance and Efficiency Through Intelligent Systems
Om Prakash Rishi (University of Kota, India) and Anukrati Sharma (University of Kota, India)
Business Science Reference • ©2017 • 269pp • H/C (ISBN: 9781522522348) • US $175.00

Maximizing Information System Availability Through Bayesian Belief Network...
Semir Ibrahimović (Bosna Bank International, Bosnia and Herzegovina) Lejla Turulja (University of Sarajevo, Bosnia and Herzegovina) and Nijaz Bajgorić (University of Sarajevo, Bosnia and Herzegovina)
Information Science Reference • ©2017 • 180pp • H/C (ISBN: 9781522522683) • US $140.00

Handbook of Research on Advanced Data Mining Techniques and Applications...
Shrawan Kumar Trivedi (BML Munjal University, India) Shubhamoy Dey (Indian Institute of Management Indore, India) Anil Kumar (BML Munjal University, India) and Tapan Kumar Panda (Jindal Global Business School, India)
Business Science Reference • © 2017 • 438pp • H/C (ISBN: 9781522520313) • US $260.00

For an enitre list of titles in this series, please visit:
https://www.igi-global.com/book-series/advances-business-information-systems-analytics/37155

IGI Global
DISSEMINATOR OF KNOWLEDGE

701 East Chocolate Avenue, Hershey, PA 17033, USA
Tel: 717-533-8845 x100 • Fax: 717-533-8661
E-Mail: cust@igi-global.com • www.igi-global.com

Table of Contents

Preface

With the increasing importance of knowledge as the basis for economic development, the issue of R&D as a source of knowledge and of innovation is increasingly under discussion. One of the major activities in support of economic development, therefore, is the setting up of a set of national support programs aimed at the support of R&D. Innovations is a complex topic. R&D management with all the topics it includes is complex by itself, and any analysis would need to be multi-disciplinary. Just the topic of Technology Transfer, one of those included in the R&D management, would involves legal issues (e.g. contracts, knowledge protection), economic issues, and organizational aspects. All that as well as technology and relevant sectorial aspects. R&D management would add to all of the above, finance, human resources, and of course management aspects; while innovation would further complicate the discussion by adding more dimensions – e.g. policy. While it is not intended to enter detailed multi-disciplinary analysis, most of the relevant issues mentioned here are introduced and discussed briefly, leaving room for further reading and study.

These R&D support programs are governmental tools used for assisting non-governmental entities in forwarding the economic development of the region or nation. They are set for long periods of time, as they support projects of long duration and require time to able to be assessed. Their increasing importance and the public funds allocated for their activities merit the discussion offered here. These programs are the focus of the discussion in this book.

The discussion will focus on the role of national and international research programs. The way they operate, the different forms they take in order to attain different targets, and the different approaches to their execution and evaluation; the special difficulty such programs have when dealing with traditional industries, and the different approaches developed. The conflict inherent in such programs involving firms and academic institutes and some ways to overcome it.

Issues related to the production and protection of the knowledge as well as the different ways to protect it are presented.

The different support programs, as well as other support tools such as accelerators and technology incubators are also presented, discussed and summed to form a general picture of the entire ecosystem.

The programs mentioned here are for illustration purposes and should be treated as such. If interested in a specific support program (to support a need in R&D) it is recommended that the reader undertake a specific search for the best fitting programs for his needs.

The purpose of the book is to give the reader a general but as complete as possible description of the R&D ecosystem (see Figure 1 of the next chapter). It includes the analysis of the R&D support programs, the different players, their goals and different points of view, as well as the tools employed.

The book is therefore, intended mainly for those interested in better understanding the components and terms of an innovation ecosystem. R&D people in the industry, seeking better understanding of the environment they are working in, as well as the other parties and their point of view; researchers in Academic organizations considering involvement with the industry, and therefore trying to gain a better understanding of the environment, which is different from the academic environment they are used to; management, innovation and especially R&D management students seeking a concise book on the topic from which to start their reading. One last but not least important audience is the general public, seeking a better understanding of the knowledge based environment that so much is heard about in the media. It is with these different audiences in mind, that the definitions and examples are presented throughout the book. The examples in particular have two main goals, the first is to further validate the arguments presented in the book; the second is to connect the theoretical sounding ideas to the real world.

The structure of the book is intended to give the reader the basic rational for the programs, and the peculiarities of their by-laws, before getting into the finer points of the different types, while also discussing, where pertinent, the supporting tools, such as policies and accelerators, technology incubators, and Technology Transfer organizations.

The discussion shall start from the basics and present the need for R&D, the R&D process and its characteristics, and some of the general introduction to the subject. Chapter 1 serves as the rational for the discussion to follow, and the main issues of the entire book. It can also be viewed as a general introduction to the complex issue of R&D and R&D support.

In Chapter 2, "Technology Creation and Technology Transfer," issues related to technology transfer, such as the need for it, its definitions and its barriers and limitations, are discussed. It aims to create the basis for further discussion regarding R&D cooperation and the relevant programs. The discussion mentions high-tech and traditional sectors as examples to issues related to technology transfer, and R&D. The different characteristics and the time element are also mentioned to illustrate some of the finer points of Technology Transfer. Also discussed are some examples to illustrate the issues.

Chapter 3, "The Unequal Partnership," discusses the players in the R&D arena, their characteristics and interests. The chapter focuses on the discussion on the unequal partnership between industry and academy when dealing with IP and its usage. It also presents the main parameters towards economic growth when dealing with knowledge creation, and market introduction of said knowledge. Among other topics discussed, are the knowledge gap between basic research and applied research, and the role of translational research. A mention of national and international research institutes is made with a full discussion left for another chapter later on.

Chapter 4, " Knowledge Packaging and Knowledge Protections," deals with some of the basics of Knowledge protection, the different options and their optimal uses. Examples help illustrate some of the points argued. The purpose of this chapter is to familiarize the reader with the basics of knowledge protections and to create the common grounds for further discussion.

Chapter 5, "R&D Support Program Types," discusses R&D support program. The program characteristics and needs as well as the different roles the different players play, are presented in this context. The description is general, followed by some of the different characteristics and types of programs. To better illustrate some of the points discussed an example of a specific sector is presented. The structure as well as the basic guidelines for establishing and R&D support program are discussed. The R&D program is defined as is translational research followed by guidelines for the operation of a program and ending with additional recommendations.

Chapter 6, "Increased Complexity in Programs," discusses the issue of complexity in R&D support programs, expanding on the previous chapter. Presented also, are the reasons for the complexity, the difference in the point of view of the government agency or the other players. The programs tend to grow more complex, from different reasons presented and discussed, as well as the advantages and disadvantages of the process. The discussion aims to be focused on the practical rather than the theoretical aspects.

Chapter 7, "Research Policy and Adjacent Tools," aims to complete the innovation support picture. The issue of Research Policy, its role and impact is presented. Following that discussion, other support tools are presented briefly - Accelerators, Technology Incubators and more.

Finally, the conclusion summarizes the entire discussion. However, before the final summary that shall sum up all the above, some of the ongoing trends are presented.

I have been involved with such programs on national and international basis for the last 25 years and have vast experience in the subject. Further to that, my PhD thesis had to do with the formation mechanisms of R&D consortia between industry and academy, a field of vast challenges, and to me of huge interest. Over the years I have seen the importance, expressed also in the public funds allocated to it, increasing, and the development of new models, better adapted to specific needs appearing.

I hope that the book will further the understanding of the R&D ecosystem and thus help overcome some of the market failures currently evident. I am sure that they will not have disappeared due to it, but some reduction in the economic and legal "friction" would contribute a lot for the general public good. It is also intended to increase the awareness for the need for increased professionalism regarding the personnel dealing with innovation, e.g. the TTU personnel, regarding their basic and further training.

The book aims to create a general but current picture, expressed also in the selection of recent literature in the bibliography and further recommended reading, over some of the more classical resources. The further recommended literature, is especially designed to offer the readers topics and resources that are current and relevant for recent practical issues, in order to further their professionalism. The book presents the multi-disciplinary topics that are involved in the world of innovation and the different disciplines that are relevant both for the practice and the research.

Chapter 1
R&D Support Programs (the Need)

ABSTRACT

Any good business plan starts with the definition of the need for which it intends to base the entire business. Following that example, the discussion here should also start with the need for such programs, their strategic role as defined by research, and the rationale that makes them prevalent. When discussing R&D support programs, one needs to bear in mind that there are different types of such programs. The specific needs such programs were designed to answer, whether well or less so, the target audience or clients they address, and the specific legal environment they operate in, influence the different types and give rise to the need for different types of programs to begin with. In this chapter, we shall define the R&D support program and analyze examples of the influence the parameters mentioned above, have on the structure and function of the programs. The different types as well as support tools will be discussed at a later chapter (Chapter 05 – R&D support program types).

THE CONCEPTUAL FRAMEWORK REGARDING R&D SUPPORT PROGRAMS

R&D support programs are a tool to promote via financial support R&D activity, with the aim of fostering increased economic development. Therefore, any discussion regarding R&D support programs shall include R&D related

DOI: 10.4018/978-1-5225-3652-9.ch001

topics, such as: R&D management, different types of research (e.g. basic. Translational, applied etc.); types of R&D performers (e.g. industrial firms, universities, research centers, Technology Transfer institutes); Technology Transfer issues (e.g. Technology Transfer process, barriers and bridges to Technology Transfer, the role of the TTO, etc.) and Intellectual Property issues (e.g. protection, licensing, etc.). This is in addition to the discussion of the added aspects the programs would have, and the basic issue of the rational for such programs and who operates them.

When discussing such a topic one should remember that we have the added burden of defining the terms each time they are discussed as R&D may mean different things when dealing with firms in different sectors. Since the topics mentioned above are interwoven it is important to present a conceptual map that will allow the discussion to progress in an orderly manner.

One should bear in mind that not all the support given to firms and other organizations can be grouped under the title of "R&D support program", and even some of the innovation schemes are different. In the subsections below we shall mark the line in the sand to differentiate between the different activities.

Figure 1 below, presents the major steps and reasons for the activities, the performers and the role of the R&D support programs as facilitators of the process. This is a very general framework, and the special cases will be mentioned when each item is discussed in turn in the book.

As can be seen in the conceptual framework. It is the responsibility or role of the firms to bring the R&D results to the market and create the economic growth. They of course have their own incentives which will be discussed below. The research organizations (in general) are involved in basic research, in the Technology Transfer to the firms and in the R&D to some extent. The R&D arena is where both research organizations and firms may be involved in. It is important to note that their involvement may vary in extent and rate from sector to sector. R&D Support programs are described in a cloud to denote that they are a collection of different types and that they sometimes include other goals in addition to direct R&D support.

A caveat should be inserted here, regarding the contribution of new knowledge from the universities to the market. There is an additional channel for the universities to forward such innovation to the market – the education system. The easiest to recognize is the via the research students, MA, MSc and PhD who following their graduation take their research capabilities and to a certain extent migrate into the industry, bringing their capabilities and knowledge to the recipient companies. To a lesser extent, it also happens with

Figure 1. R&D process conceptual framework

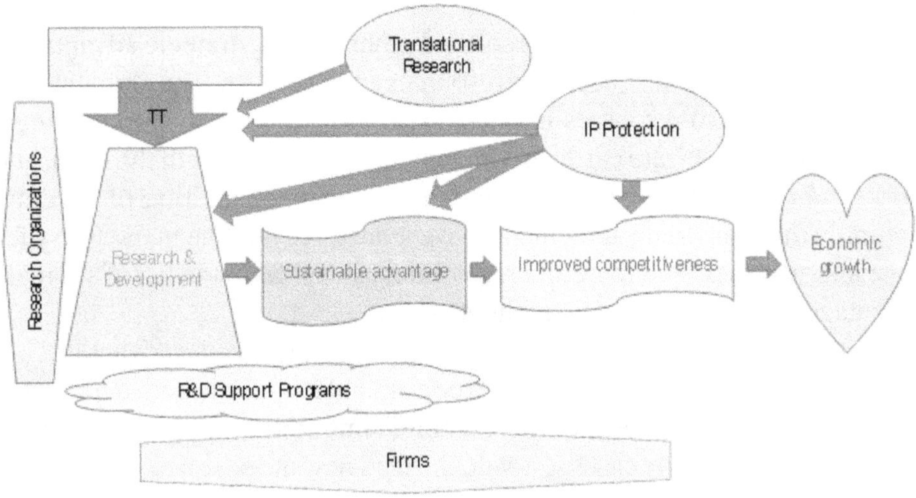

the BA and BSc graduates. Some programs support the migration of similar HR to the industry (e.g. Marie-Curie programs in Horizon 2020).

THE NEED

The increased prevalence of international trade agreements opening local markets to international competition, increased the rate of globalization. The globalization of world markets is making competition ever more intense. In most production sectors, this is best evidenced by the prevalence of the "Made in China" trademark. An interesting example is the fast disappearance of PV manufacturers in many local markets, in favor of Chinese manufacturers.

This has also affected less technology enhanced sectors such as the publication sector. Today, as it is so easy to publish your own work, it is becoming more and more difficult, not to publish your work, but to get anyone to notice it. Every trick, advantage in the market is a better term, is welcome as it allows the publisher to get his publication noticed and gain his income from it. Bloggers are rated according to the number of followers they have.

For firms in a global market, where there is competition coming from all over the world, some tangible and strong advantage in the market is required. As Michael Porter (1980, 1982, 1985) already in the 1980's and others have claimed, knowledge is the future source of advantage. The EU and others have recognized this and have made the Knowledge Based Economy its target (UNESCO, 2005). It has mostly replaced other former strategic advantages, such as access to markets, access to resources and expertise. The globalization has made access to resources in many cases easy enough to stop it being a strategic advantage (rare earths in China being the exclusion to the rule), and worker's mobility has made the access to expertise a question of funding. Naturally the same trade agreements have made access to the market easy for most internationals and therefore removed it as well from the list of strategic advantages, except in specific niche markets.

It would seem that Knowledge forms a good solid strategic advantage, but strangely enough it is also viewed as a consumable, which is both degradable and requires constant investment in updating. Time and new advances make the old knowledge, in many cases less valuable and in some obsolete. Imagine the knowledge embedded in the original IBM PC of the 1980's relevance today. In fact, most of it was already obsolete ten years after it came on the market. If compared to older capabilities, the ability to correctly use the hammer to hit the nail entailed a tool (the hammer) as well as the knowledge how to use it. Both have been relevant hundreds of years ago and are still relevant. However, the 1980's IBM PC's are no longer relevant as tools, but so are the skills required to use their now obsolete operation system. Therefore, as the technology advances fast, there is a need to renew the knowledge behind it.

The process of renewing the knowledge which we wish to discuss here is the R&D process.

THE RATIONALE BEHIND R&D SUPPORT PROGRAMS

As mentioned above the competitiveness in the markets except in very rare occasions is increasing. With the international trade agreements opening the markets more and more to foreign entries, reducing tariffs and customs and the introduction of globalization in consumers and related market, companies find their old markets more crowded and need to find new means of attracting the customers and maintaining both their market share and profit. In a more aggressive outlook, the same process of opening markets is presenting firms with opportunities, to which they need to be prepared, the entry into new markets is not easy for large and efficient firms, and is more difficult for small and resource limited firms. In the last decades, a new term has emerged as the "silver bullet" or common explanation for success and the lack of it as the common excuse for failure – “Innovation”. From where can that innovation come to the firms? What is that innovation?

Degrees of Innovations: The Place of R&D

Innovation is not limited to the technological aspect of the firms, either in the products or in the manufacturing; it extends into other areas of operation, and has long since migrated into noncommercial fields of activity in our lives. The term refers to the ability to do or offer something in a way it had not been done before.

- Very innovative activities are such that have never been done before by anyone. Social networking is such an activity.
- A lesser degree of innovation is when the action has not been done in that sector or for that use before, but it is known or practiced in some other sector or for another purpose. The new use of a known drug can be such an innovation.
- The last innovative action is one that has been done by others but never before by the specific firm – localized innovation.

Looking at these three degrees of innovation one can see that the lesser two forms of innovation can be attained by observing others; either by observing other firms in your sector, or looking broadly at other firms in other sectors and importing their innovation.

But the higher degree of innovation can be attained only by creating something new that has never been done before, inventing something. The main process of invention from the technological point of view, is the R&D process which allows firms to create new knowledge owned by them - the ownership point is paramount and a separate chapter is dedicated to it. At this point we shall just mention that the ownership of the invention is important as it allows the firm to create a temporary monopoly in the market, a long-term advantage in the market, which in the above-mentioned competitiveness situation is crucial to the firm’s success. It is that reason which makes the higher innovation the most interesting of the three, as it allows the creation of competitive advantage in the market while the lowest is mostly keep up with the competition.

This qualitative argument is supported by quantitative research as well. Rosenberg (1990) showed that R&D is beneficial to the firms engaged in such an activity and Porath (2010) showed that Firms dealing with generic

R&D perform better in in the market in times of crisis. Knowledge has been acknowledged as a source of economic growth in the world as is evidenced by the EU decision to base the EU economy on what it chose to call the Knowledge Based Society and initiated the Barcelona Initiative with its R&D goals, of 3% of GDP invested in R&D and at least 66% of the funds coming from the private sector (EU, 1995).

The rationale is that R&D allows firms to develop advantages in the market, either by superior product, improved quality, better production process or managerial aspects including finance and marketing (Rosenberg 1990). These advantages in the market allow the firms to increase their competitiveness and thus create more income, and economic growth that is also expressed in increased employment. So, it is clear why firms chose to engage in such an activity, it will give them an advantage in the market and increase their income. That would justify the financial investment, or more accurately the allocation of resources firms invest in R&D – financial, HR, and management time and attention as the major resources invested.

Accountability

An important aspect of the programs that we discuss in this book is that the funding for such program, at least in part is coming from public funds.

Therefore, an important aspect of such programs is the need to justify and to account for the support given by them. It has a more profound impact on the structure and by-laws of such programs as we shall see below when dealing with theses aspect, the point at this time to remember, that the justification for existence as well as the continuance of activity for such programs depends on the accountability of the management of public funds.

The improved performance during times of crisis could justify the decision to invest as strategic, from the relevant firm's point of view, but why would governments invest? Why should Governments, national or local, invest in the creation of financial schemes that drain their public funds?

Economic growth in many cases is difficult to measure directly. If you have a firm, active in the market, selling products and services in several market segments, it is difficult to distinguish the improved performance of the firm as stemming from a single source – in our case the R&D activity. The economic growth could be attributed to a general growth in the world market, the local market, the specific sector, the improved financial terms, improved management capabilities, better marketing etc. or in most cases all of them together. Therefore, it is important to create a measurable indicator for the effectiveness of the program, preferably one that is not wholly based on the general satisfaction of the applicants – they may be happy just to get the funds. The simplest indicator would for the creation of start-ups, as they go from zero activity to something, and the entire delta can be credited to the supporting program.

In addition to the benefits to exiting firms, there are also additional benefits for R&D, especially the R&D that leads to the creation of new firms and thus increases the economic growth. Of the two advantages, to existing firms on one hand and the creation of new firms on the other, the easiest to quantify is the formation of new companies which is more easily measured. However, existing (veteran) firms have less chances of failure and they have a better understanding of market needs.

One important aspect, especially for the discussion here, further down, is that the R&D performed by firms, supported or not by the R&D support programs, normally leads to the first benefit, as the firms would retain the knowledge and the resulting advantage for themselves, only rarely does it lead to the formation of new firms. It is more common for R&D performed by research organizations, to lead to the creation of new firms, based on the research results. There is one major exception to that rule and that is the formation of a firm based on an idea and performing the research later on, in which case the new firm's main activity is often the R&D project alone. However, this, the case of the technology start-up, could be seen as a private case of the R&D performed by firms.

Therefore, from the government point of view, the allocation of public funds to R&D support programs is viewed as an investment in increasing the potential for economic growth, expressed over two plains, assistance for the support and well-being of existing firms, increased economic gains, and increased employment, as well as increased employment and economic growth derived from the creation of new firms. The basis for the creation of the financial tools is therefore socio-economic in nature (Lach, Parizat and Wasserteil, 2008).

THE R&D AND R&D SUPPORT PROGRAM

The discussion so far has been very general regarding the R&D and the support programs, but very specific regarding certain qualities and characteristics. We need however to define some terms before the discussion proceeds.

R&D Process

Research and development are in fact two actions, along the knowledge creation path, leading from the derive to better understand nature (Research) towards the appearance of a specific application that can be utilized commercially (Development). The process is in most cases lengthy, performed by different entities along the line and therefore requires knowledge development as well as Technology Transfer, which shall be discussed in the following chapter.

It is important to note that for the purpose of discussion in this book we refer to R&D process for technologies leading into products, process or a combination of both. Social development or other activities devoid of direct technological development are not included in the discussion. Therefore, for our purposes the redesign (e.g. new color) of a product, that does not include a new or improved performance or application is not considered as an R&D activity, and while it may entail effort and cost, and result in improved sales, is not part of the discussion here.

R&D Support Program

Following the explanation above regarding the nature of the R&D process for our purposes, it should be clear that the relevant support programs are those that support the process or a specific part of it.

The R&D support program will support the knowledge creation and validation activities as well as the required knowledge transfer and the enhancement of R&D capabilities of the applicants or their personnel including the inception of additional workers. It will not support activities such as marketing publications, etc. in most cases, as well as other "innovation activities" that do not refer directly to technological R&D.

Therefore, an innovation activity that is intended to allow the applicants to computerize their accounting and billing systems – utilizing existing systems sold on the market, while it may well modernize the performance of the firms and may well have a positive impact on the performance of the firm and its economic growth, is not part of the activity here, even if it is a move supported by a public entity.

Some innovation activities that are aimed at improved performance from the marketing point of view may not be included, even when they are supported by a local economic development agency or similar. It is important to note that for firms who incorporate such software for accounting or that improve their marketing, as we mentioned above – this is not an R&D activity and therefore the programs supporting them are not considered here, however for the companies developing such software or technological tools for improved marketing, the development of such software and tools, may well be an R&D activity and therefore support programs for such activities are included in our discussion.

Example 1: Industiral Design

The Israel Small and Medium Enterprises Authority (ISMEA) supports Small and medium size companies in an "industrial product design" activity by offering a support program. Such an activity while being clearly for the benefit of the firms and their economic growth and an innovation activity is not an R&D activity and the program is not an R&D support program (Israel Business Connection, n.d.).

Academic funds shall be included as a sub-set case of R&D support programs, as they refer to a specific sector performing normally only the basic research activity. However, it would seem that to ignore the beginning of the process is wrong, while not referring to sideline activities may be more productive.

The national academic funds therefore shall be regarded as R&D support programs, while the internal research funds of such organizations shall be treated like the internal funds firm utilize for development projects inside the firms.

Unlike the academic funds, Direct Inward Investment efforts are not discussed in this book, as they are not directly involved in the R&D effort but rather support the economic involvement of foreign entities with the intent of assisting economic growth in this way.

Other Tools

It should be mentioned at this point that there may be other support tools than just R&D support programs (e.g. Direct investment support programs). There can be, additionally, institutes that allow for the transfer of basic knowledge to firms, or even directly to the market. These institutes will work jointly with the industry and help it develop its products. Such institutes require a heavy investment and on-going maintenance and are therefore an alternative or additional tool to the R&D support programs.

Similarly, there are accelerators, technology incubators and VC networks (including Angels) that assist in the spin-off process, and help create new start-up companies to bring the Research results to the market. These will be briefly discussed below.

GEOGRAPHICAL/NATIONAL ASPECTS

The programs mentioned above have been defined geographically as regional. The region can be an entire country or smaller regions within a country, or even a combination of several national entities. The regional definition stems from a combination of legal, budgetary and political demands.

As mentioned above an R&D Support program would require the usage of public funds. In most systems that is more easily done within the same political/geographical unit. The usage of national funds for foreign economic development would sound very problematic for any entity that would need to approve them.

From historical reasons, the programs stem from strategic cooperation aimed at increasing the competition of a specific geographical unit against the competition from the rest of the world.

Regional programs depend on two aspects, the ability of the region to divert funds to its economic development – normally relevant for "weaker" regions requiring help in their economic development, and a relevant agency to promote execute and monitor the activity. The EU is an example where the usage of such programs as economic tools was considered expedient towards the further unification of the EU, from the economic point of view. The EU is a special case in that, the economic unity preceded the political or even social unity[1]. It is further interesting to note that the participation even though it started on a geographical basis acquires some of the strategic political aspects – see the EUREKA and Horizon 2020 participating countries lists[2]. These lists include countries that are members of the EU, but also countries that seem important strategically such as other non-EU member European countries, as well as China, India, Brazil and additional countries, with scientific or technological ties to the EU.

When dealing with regions in a specific country, whether they are economically independent or relaying wholly on support from central government, it is important to note how the goals of the support program are defined so as to promote the specific regional agenda. Such programs normally intend to enhance existing capabilities in the specific region, building on the strengths existing and trying to prevent the deterioration of these capabilities.

As we discussed above, there are programs for local support for economic growth that are not R&D inclined, and while these are just as important – they are not part of the discussion here.

LEGAL ASPECTS

The legal aspects regarding R&D support programs can be divided into two major groups. The first one is the legal aspects regarding the definition, approval and execution and monitoring of the programs themselves; the second refers to the IP rules relevant in the domain in which the programs are active.

In this chapter only the first type will be discussed, as the second type has a chapter all to itself below (see Chapter 04 - Knowledge packaging and Knowledge protections.).

When discussing R&D Support programs one should bear in mind that this is a case of government interference with free market forces. Furthermore, it involves the allocation of public funds to (in most cases) private organizations that are profit oriented, and that indeed in many cases the preference for allocation is the ability of the public funds recipient to turn that support into an economic and financial gain. We are so used to that activity that we rarely stop to think that if in any other case public funds were allocated to private organizations, so that they can profit, a public debate regarding the morality involved, as well as the allocation of public funds for the benefit of private persons would be entered into.

Is this really a subsidy to private organizations that is being distributed? The answer is more complicated than can be discerned at first glance.

If a company is performing its R&D (anyway, with or without the support) than this is definitely a subsidy that reduces the costs incurred by the company, and helps the company reduce risks, increase the profit as well as other economic direct gains from the activity itself. Considering the Spill-Over effect of Knowledge, the funds are necessary in order for the activity to begin, or proceed, either in academic research which would not happen without the funds, or a start-up company that could not do the program without the fund, even if it participates in the funding itself – than this is not really a subsidy but rather an enabling activity (Luukkonen, 1998; Luukkonen 2000; Porath, 2010).

In many cases however, the exact crucial role of the funding is not the only part of the assessment but rather the form in which it is supplied. If the funding is supplied as grants (non – returnable, no dividends attached), then one may assume that this is subsidy unless like mentioned above it is the crucial support enabling the entire activity – SME support, academic research and similar activities or combinations of the above.

On the other hand, if the funds are supplied in the form of a loan, that is reimbursed from sales of the R&D results and derivatives is required until the funding plus interest is returned, then this again is not a real subsidy.

The support for single firms is a simpler issue, the support to a group of firms in a certain region is another issue as the question of cartelization is raised.

Cases of cooperation between firms especially in the same sectors of industry, R&D consortia, could be suspect of cartelization. However, as well all know, that is not the case. There are several customs that have become accepted generally, that regulate the support in a way that would seem to make it fairer in general and that would allow what is considered a beneficial activity to go on; these rules address two main aspects: selection and support.

Selection Rules

- The support is given on a fair and equal basis to all relevant entities in a predefined region wishing to request it;
- It is publicly published beforehand allowing all relevant competitors to apply;
- The rules for selection are predefined and published;
- The activity is determined to advance causes for the general benefit (economic growth is considered to be for general benefit);
- The programs are subject to the same scrutiny all similar government activities are.

Support Rules

- Generally, the support is given as a loan[3]. If the project is successful and results in economic gains to the recipient of the loan, the funds plus index, interest or both have to be returned to the funding government. This way the unfair advantage due to the support is someone negated by adding a sort of "fine", and the public get a direct benefit as more funds are made available based on the success. If the project fails to result in gain, the public bears the risk;

- Funds are supplied on a participatory basis and their size is capped by predefined regulations, as are the expanses they are allocated for; projects are monitored both technically and financially;
- General information about the utilization of the funds is published, and is part of the scrutiny of similar government activities.

FINANCIAL/BUDGETARY ASPECTS

Some of the budgetary aspects were discussed above as the different aspects are interlinked and cannot really be discussed completely separately. However, in this section we need to get into some of the finer details.

The support for an R&D project does not necessarily require full financial support. Having said that, there are very few programs that do not involve at least partial financial support, as the R&D activity is cost incurring, but there are additional components required.

As Luukkonen (1980, 2000) showed, the role of R&D support programs and especially national or ultra-national programs is to reduce the risks for firms, in order for them to pursue projects that otherwise would not pursued as they would be deemed too risky. A significant part of the risk for firms at least, is the financial aspect. It is important to remember that for firms, and especially for firm management the main problem is not always lack of funds, but rather the allocation of the funds among the different tasks the firm has to perform.

Fund allocation often includes aspects of risk analysis and risk mitigation, where the management has to justify the decisions made regarding the potential benefit resulting from these actions, versus other potential activities. Therefore, by reducing the risks and the amount of funding required for the R&D projects, they become more attractive for the firms and the chance of the firms proposing such projects would increase. The need for riskier projects is derived from the growing demand for disruptive technologies, and it can be easier to assume that the disruptive technologies would come from riskier projects than from mainstream R&D.

Having explained the main rational for the R&D financial support component in the projects, since as we mentioned above the funding comes from public funds, it requires among other aspects strict and clear regulations as to the allocation, delivery and monitoring and reporting of the expanses as well as the terms under which the finance is done.

Types of Finance

We can divide the financial support given into two categories regarding the rate of finance.

- Full reimbursement – the financial support can cover the entire cost of the project, thus reducing the financial risk of the executing organization to zero. In some cases, the funding is supplied before the project commences. This type of financial aid is typical of research support funds for the academy supporting basic research. In such cases the main output of research results is a scientific publication, and as commercialization of the research results is relatively remote, the funds supporting that kind of research seek the prestige and mention in the public as sign of their effectiveness. Since academic research is performed by non-for-profit organizations in most cases, and they, furthermore, do not compete in the market, accountability is considered less important. When this is done, there are some aspects that need to be taken into account:
- Partial reimbursement – in which the funding supports a certain percentage of the approved budget for the project. Therefore, we have here an approved budget for the project, and reports of the actual usage of that budget, against which reports the financial support is given. In such cases the approved budget represents not the actual cost of the project, but the maximum funding available for the project. Due to the nature of such projects, reporting both of progress or if the project duration is very short, at least summary reports are required. These reports need to be audited, either by the support program's administration or by external auditors and the financial support transfer to the awarded entity.

  In cases where such project would burden the awarded entity with intermediate financing the actual support may be divided into the following kinds of payments:
  - Preliminary payment (advance) on account of the approved budget – normally would be between 20-33%, depending on the next period for reporting whether it is quarterly, trimestral or semiannual.
  - Progress (interim) payments, taking into account the progress of the project measured by the approved expanses

of the project, and deducting the advance – not always there, depending on the length of project period.

- Final payment – the settling of the payments due to the executor of the project under the approved budget.

This division of the type of financial support is not the only possible categorization. Another categorization is regarding the conditions under which the financial support is given. The following conditions apply in most cases:

- Grant – the financing is transferred to the executor of the project and no returns or reimbursement are expected, unless the project is not executed for some reason. Grants characterize financial support given to academic research, for the extension of human knowledge. But it is not limited to academic research alone.
- Loan – the financial support is given to the executor of the project, but that finance is to be reimbursed under certain conditions and the funds returned to the program administration. This is normally the case when dealing with product development, and the reimbursement is to come from sales of the product which the project helped develop.

There are of course combinations of the categories mentioned above that are also available. The funding may be supplied as a loan, but if the project fails or no sales are generated, the loan is converted into a grant. There are also possibilities of combination regarding the conditions differentiated according to the type of the project executor, the financing maybe in the form of grant for an academic entity and in the form of a loan to a company.

There can also be combinations between the first division and the financial conditions. There may be a single payment of a grant fully covering the project cost, or a partial support of a project under loan conditions or other potential combinations.

- The determining test regarding the type and conditions of financial support are normally
    - Academy as per its reduced ability to commercialize the results - will be granted either to a full financial support or close to it;
    - Industry and national research organizations selling their services, would normally be able to perform projects even when supported only partially as well as under full support, and also be able to do

so under loan conditions as they expect income to result from the R&D effort.

- The closeness of the results to the market:
    - When the results of the project are a product ready to be sold the tendency to insist of partial support as well as loan conditions are stronger, also in order to allow for the involvement to assist the market and not create unfair competition;
    - When dealing either with research or with technological development and not with product development the demand for reimbursement from sales is complicated to execute as well as track the specific support contribution to the final product sold on the market.
- Legal anti-collusion and government support regulations of the specific country or region.

All of the above refers to the conditions of financial support as seen in the two-way lane between the executing organization and the support program. There can of course be projects where different partners are supported under different conditions for example, as per their identity. At the same type, some programs may demand that one type of partners be supported by the others – e.g. academic research organizations be supported by the industrial partners in a program where the support to the academy is partial. In such cases the industry will complement the support given to the academy to 100% or close enough for the academy to be able to execute their part. Similar cases may apply to start-ups or other organizations that cannot perform their part without the full financial support.

We have not discussed the sums actually supplied as this vary greatly regarding the type of R&D performed, the specific country and cost of living and HR and other aspects. However, we can discuss the type of expanses that are recognized as eligible for support:

- HR – financial support to cover the expanses of the personnel involved in the work are normally a basic and approved cost. It is of course impossible to execute any R&D project without any involvement of personnel (we discuss sub-contractors next), and therefore this cost category is paramount in importance. The caveat one finds from time to time that permanent personnel, especially in governmental organizations, cannot in many cases be supported by additional funds. But even in such cases, students, additional personnel incurred for the projects and similar are supported. Another aspect in that regard, is the report of Men Months or hours that is used to estimate the effort invested in the projects. Assuming that the time people invest is more universal than the salary they receive as an indicator of the R&D effort;

- Sub-contractors – in many cases these are a supplement to the HR, either by supplying activities or expertise that the executing organization does not own, or want to own and is therefore hiring temporarily for the specific project task. Sub-contractors are required to perform specific pre-defined and approved tasks. In that regard, in most cases sub-contracting is referred to as a black box and therefore the time invested or the specific salary of the personnel involved from the sub-contractor's side is not measured. Some programs, due to that aspect of the black box limit the relative size of the sub-contracting to a certain percentage of the overall project budget. Most programs limit the sub-contractor category in order to assure the real participation of the applicant;
- Consumables – most projects would utilize materials or other consumables, which by definition would be items used for the project during the project duration which cannot be used again after the project has ended. Materials used in experiments, mice, electrical components etc. are examples of such consumables. Of course, the usage of consumables is monitored and normally only what is consumed during the project is reimbursed within reason;
- Equipment – items that can be used during the project execution, but also remain functional as a whole and can be used after the project has terminated. A good example is a computer, which would be used during the project life, but can also continue to be useful and used after the project has terminated. Due to the fact that the equipment last after the project terminates there are several main methodologies in dealing with equipment costs:
    - Full cost reimbursement – the equipment is critical for the project and the funding entity does not want make this specific cost complicated in reporting or monitoring and therefore full reimbursement is given based on evidence of purchase;
    - Depreciation – in order to account for the usage of the equipment in the project, the program would recognize the depreciation in the equipment over the period it was used in the project. This is normally in conformity of the depreciation regulation in force in the legal environment of the program;

    - No recognition of any equipment costs as the main benefit from the equipment can be derived for many years by the executing entity and therefore it should find other means for such support.
- IP protection – the cost of applications of different types in order to assure that the knowledge created in the project is protected and ready for commercialization;
- Software licensing – software licensing is not consumed neither is it a depreciable equipment and therefore in many cases is a category by itself;
- Marketing – this direct cost is not always included in the budget consideration of an R&D support program, and even if it is, it would normally be considered only regarding marketing surveys and familiarity with the market, but only rarely more than that. Market survey can be viewed as part of the development process, a required information in order to direct the development to respond and fit the market demands regarding characterization of the product. As well as art of the assessment of the market potential of the product, which may be important for assessing the value for the firm/ project and thus impacting on the ability to raise further funds from government as well as from private sources. Of course, this is more crucial for Start-Ups and firms not active in the market, while firms already involved in the markets would in most cases either have access to such knowledge, or in the best cases start from such market demand.
- Overhead – an indirect type of cost. The costs of the R&D project, the direct costs do not stand alone, from accompanying costs such as electricity municipal tasks, to HR recruiting costs and other costs, there are indirect costs that accompany every project. The ability of attributing the exact indirect costs to the specific project is often limited and even when possible would often increase the costs. Therefore, in many programs it is considered more cost beneficial to determine a fixed rate for the overhead, which is normally indexed to at least one of the direct costs types – either as a percentage of the whole budget, the HR costs or a combination of these possibilities. Of course, there is a possibility to demand an assessment of the indirect costs but as mentioned above that would involve tracking and attributing each overhead type cost to the project and later on require the governing agency to monitor and audit these cost items.

STRATEGIC ASPECTS

The strategic aspects of the R&D Support program for the agency governing them and for the participators are more complex than may appear at first. When mentioning strategic aspects, the intention is to deal with long term goals with a wide impact. Regional economic growth would be such a strategic aspect, the formation of a single company would probably not be regarded as such from the point of view of the governing agency. From the point of view of the academy the licensing of a single technology would not necessarily be strategic, but the creation of relational assets with the industry that could lead to future joint projects and deals, would be strategic. From the industrial point of view, the ability to create mechanisms known to the industry for the receipt of knowledge from the academy, or to create the specialty for future cooperation would be strategic while the single project in most cases would not be considered strategic.

The Governing Agency

The obvious strategic aspect is the economic growth, and that of course exists. The direct growth from the activity is normally accompanied by indirect growth, resulting from additional economic activity attracted to the region due to its activity, and its potential benefit for new comers. Therefore, the direct benefit should not be forgotten but the secondary affect should be remembered as well.

The governing agency is also acquiring a reputation, and may find future ventures it offers better received, and its position in the competition on limited resources improved due to that activity. The secondary impact such an activity may have should not be ignored. If the concept of a knowledge based society is accepted, it is important for agencies promoting economic growth based on knowledge to understand that they have a crucial role in demonstrating the ability to lead and manage such activity.

In the section below we shall delve into the specific roles of the different participants, but at this point it is important to stress the role of the governing agency.

As a bank should demonstrate its ability to manage finances well, the governing agency needs to establish its ability to manage the R&D Support programs well, in order for the other players to agree to participate, and in order for the economic magnet, attracting the growth to function. Inability, would create disappointment and lead to the other players leaving the game, seeking similar rewards in other places. Or even deciding that the participation in such activities is not for them – a situation much more difficult to rectify.

Academic institutes which seek programs in other places, may agree to try again if the agency is replaced or reorganized, and except for the time and certain opportunities lost the actual long-term damage is small. However, if said organizations decide that participation in such programs is not for them, that they shall deal only with basic research and disdain any contact with the industry that is more difficult to rectify.

Therefore, the agency needs to demonstrate the ability to create projects, on an on-going regular basis, that fit the definitions set out in the programs. It should therefore make sure that each of the other participants gets some required, and important enough benefit from it, and that major hurdles towards participations are removed. The agency should care about the needs and apprehensions of the other participants, because as we shall discuss in this book, money alone does not answer all the needs. In the subsections below, we discuss the different participants, each with its own point of view and needs, and the potential obstacles for cooperation or participation in and R&D support program.

The Participants

In this section, we shall look at the main players and their role in the R&D support programs. We shall focus on their roles in the programs and not in general in society. It is important to recall that the focus on these roles may give a wrong impression regarding the importance this activity holds in the strategic goals of the players.

Academy

The academy has the major role of the original research and knowledge creation. It is responsible for the creation and discovery of the basic knowledge on which the research is based.

The academy performs research, more accurately, basic research which is curiosity driven, seeking a better understanding of the world and the way things work.

That research, many times of international quality and high level is however conducted under different conditions than the industry would conduct their development.

It is important to note that while the process of research and that of development may look similar to external observer, they are in fact very different.

Both research and development can be arduous, conducted according to specific and details procedures and involve innovation. However, while research is conducted under academic guidelines and is therefore aimed at the generation of new knowledge, development is aimed at the creation of a new product service or process that will be marketable. Therefore, while research is less focused on the end product and more focused on the methodology and the ability to identify the parameters and answering the research question, it can lead, by the results it obtains to development.

Academy, dealing with research is therefore less conscious of market issues, regulatory issues and standards regarding products and processes. It is also less conscious of commercial aspects regarding the research results and also regarding the processes and methodologies employed. If a component for the research needs to be coated with gold, the researchers will do it, the fact that no consumer product could be the same, as it would make it too expansive is not relevant for the research.

Academic institutes have been evolving since the medieval times for the purpose of preserving and transferring knowledge to future generations, and from the renaissance for the creation of new knowledge. They have evolved a culture that praises quality work, low salaries, and above all international acclaim. Cooperation and exchange of knowledge has long existed and had flourished and supported by publication – the main means of communicating knowledge among the researchers and also an evaluation tool of their work.

Academic research is therefore different by nature from that of the industry.

Industry

The industry is a global term referring to economic entities involved in the production and sale of process and products either to other economic entities, to people, or both. In most cases the when discussed in this book the industry described here is performing development to some extent and knows how to perform it. Whenever the discussion refers to companies that do not perform development or perform it rarely or at a very low intensity it will be mentioned specifically.

The main relevant attribute of the industry as mentioned here is that it has a presence in a market and is aimed at the generation of revenues and profits.

The basic assumption is that firms and industry are the main source of employment and of the creation of economic growth, by their production and manufacturing and that R&D activity, mostly development is an activity they undertake in order to create and maintain some strategic advantage in the market segment, or segments they are active in, in order to allow them to generate the income they require.

It is therefore assumed that the R&D activity is a tool, a mean, and not an end for the firms. They do not do the R&D activity for the economic benefit of the grant the R&D support program offers, but rather that this grant is a subsidy for the R&D activity they would undertake for strategic reasons.

Others

Since this is not a compendium of all the entities involved in R&D activities around the world, the list of types is also limited to the main types that are evidently dealing with the R&D activity in a significant intensity.

In addition to the Academy and Industry mentioned above there are research organizations, on national, regional or other level that are normally public legal entities and that play a role in the R&D activity which is partly covered in other places by the industry and the academy. They serve as a sort of go-between, and have created a specialization in that area.

Such research organizations (e.g. Fraunhofer, TNO and others) are normally supported at least partly by the relevant government and are offering their services to the industry for a fee. They perform translational research, translating the research and research results from academic research into development raw materials for the development process of new products.

As mentioned above such organizations have developed an expertise in that they can work with the academy on a certain level, taking the results of the research and translating them into a development process that can either result in a product (including a production file), that the organization will transfer to the industry, being able to understand the needs and demand of the industry as well.

Unlike the academy, they do not have the need to publish or the academic culture that has evolved in the universities, and they are also exempt from normal teaching, even if some of them do offer advanced degrees' students the possibilities of participating in research (e.g. the Pasteur Institute).

They are therefore not academic in that sense of the word, but are a specialized entity in the middle between the two cultures.

The advantage of such organizations is clear; however, they entail a certain cost to establish and maintain and therefore are not as prevalent as logic may suggest. They will be further discussed below.

These organizations however are not the only possible additional participators. There are different third sector organizations, mostly associations and companies for the public good who may be involved in projects where the target users of the final results are their constituents. For example, an association for the deaf may be involved in research regarding electronic ears, speech to written translators etc. and therefore may be involved.

Other potential participants may be companies that are not normally involved in R&D, such as in traditional sectors or even commerce and who would be involved in the projects, but in most such cases not directly performing any of the research work, rather utilizing one of the research organizations, universities colleges or regional research organizations to perform the actual research for them. In some cases, as mentioned one may find that an importer of certain goods, may try to reduce costs by asking a knowledge provider to develop a production process for him. Different government entities have long been interested in R&D for their own needs, but unless it is research to support policy, and especially when application requires some production, they prefer to have one of the other participants (industry or academy and industry) perform it for them.

The diversification of organizations having some R&D experience will continue to grow as more and more organizations, realize that they need to innovate and then R&D is a promising path towards innovation.

The preference of most of the different types that perform R&D occasionally or rarely, is to have either a knowledge provider or a firm perform it for them. Therefore, we have focused our analysis mainly of the types mentioned above, but as we already said, this may change in the future.

REFERENCES

EU. (1995). Barcelona Declaration. Adopted at the Euro-Mediterranean Conference, November 27-28. Retrieved June 2017 from https://www.cvce.eu/en/obj/barcelona_declaration_27_and_28_november_1995-en-0beb3332-0bba-4d00-bd07-46d8f758d897.html

Israel Business Connection. (n. d.). The Israel Small and Medium Enterprises Authority (ISMEA). Retrieved from http://www.israelbusiness.org.il/startingyourbusiness/assistingcenters/smallbusiness

Lach, S., Parizat, S., & Wasserteil, D. (2008). The impact of government support to industrial R&D on the Israeli economy (Final report translated from Hebrew). E.G.P Applied Economics Ltd., Israel

Luukkonen, T. (1998). The difficulties in assessing the impact of EU framework programs. *Research Policy*, *27*(6), 599–610. doi:10.1016/S0048-7333(98)00058-4

Luukkonen, T. (2000). Additionality of EU framework programs. *Research Policy*, *29*(6), 711–724. doi:10.1016/S0048-7333(99)00041-4

Porath, A. (2010). *Collaborative Research in Management: Inside out*. UK: SAGE Publications.

Porter, M. E. (1980). *Competitive strategy: techniques for analyzing industries and competitors*. New York, NY, USA: The Free Press.

Porter, M. E. (1982). *Cases in Competitive Strategy*. New York, NY, USA: The Free Press.

Porter, M. E. (1985). *Competitive Advantage: Creating and Sustaining Superior Performance*. New York, NY, USA: The Free Press.

Rosenberg, N. (1990). Why do firms do basic research (with their own money)? *Research Policy*, *19*(2), 165–174. doi:10.1016/0048-7333(90)90046-9

UNESCO. (2005). Towards a knowledge based society. Retrieved May 14 2017 from http://unesdoc.unesco.org/images/0014/001418/141843e.pdf

ENDNOTES

1 Still requiring some work towards completion.

2 http://www.eurekanetwork.org/https://ec.europa.eu/programmes/horizon2020/.

3 There are rare cases of grants, normally when the support is for an R&D activity that does not lead directly to a product, but rather to a technological platform development, that will allow the firms to continue (based on separate funding) to product development.

Chapter 2
Technology Creation and Technology Transfer

ABSTRACT

In this chapter issues related to Technology Transfer, such as the need for it, its definitions and its barriers and limitations, are discussed. The Chapter aims to create the basis for further discussion regarding R&D cooperation and the relevant programs. The discussion mentions high-tech and traditional sectors as examples to issues related to Technology Transfer, and R&D. The different characteristics and the time element are also mentioned to illustrate some of the finer points of Technology Transfer. The chapter also presents some examples to illustrate the issues.

CHARACTERISTICS

We can discuss innovation on several levels. The easiest division is between technological innovation and non-technological innovation.

Non-technological innovation would include design, introduction of improved marketing strategies and actions and similar aspects. It can also be referred to as innovation not including the creation of new knowledge. One of the important aspects of that type of innovation is that in many cases it is innovation intrinsic to the company or organization implementing it. As it does not involve the creation of new knowledge it is automatically based on existing knowledge, even if not at the level of the company, or sector.

DOI: 10.4018/978-1-5225-3652-9.ch002

Another type of innovation is innovation based on knowledge creation. The knowledge creation can be technological or it can be based on technology developed by others. When the new knowledge is created, either by the firm implementing it, or by someone else, it can be protected (see Chapter 04—Knowledge packaging and knowledge protections) and can be used for dealing – selling and buying, transferred and divided, or added to.

External innovation includes Knowledge imported from the outside of the firm, either from the public domain or from other entities and implemented inside the firm. The introduction of an ERP system to improve the firm performance, compliance with standards, or a purchased patent can be additional examples. External innovation requires openness from the firm receiving it and can also be termed "Open Innovation".

When internal innovation is encountered there is very little complication as we are dealing with a single entity doing the work, and enjoying the results of the work. When such internal innovation involves external funding, it may come from an R&D support program. In many cases the firms involved in internal development can choose to do so based on their own funds.

When dealing with Open Innovation, involving external sources and the firm, there are several organizations involved in addition to the funding source, and although there are instances when each of the participants fund their own activity it is very common for them to prefer an R&D Support Program, especially when the partners are not equal partners either in size or in marketing abilities. Since firms financing their own activities can be viewed as a private matter similar to the self-financing of a single firm and therefore not part of the discussion in this book. In this book, we shall discuss the R&D support program, and will differentiate between the programs supporting a single entity and several entities.

One of the most prevalent terms when dealing with Open Innovation as mentioned above, is that Innovation coming at least in part from outside the organization, is Technology Transfer. It is a major aspect when dealing with most of the R&D Support programs, as in any case when there is more than one entity performing the R&D, Technology Transfer, at least in part is involved.

Technology Transfer is a transfer of the technological knowledge or part of it from one entity to another. It is done in order to allow the receiving entity to be able to acquire a required technology either for direct commercial use, or in order to further develop the technology towards that end. It is important to note that the transfer of the technology does not need to happen serially, but can be concurrent.

An example would seem to be in order here.

Technology Transfer: Example 1

A firm producing and selling refrigeration units based on solar panels is interested in improving its cooling timespan by adding a storage capacity to its products (chargeable battery). It contacts a research entity (university) that develops storage capacity technologies, and the university offers the company a prototype it has developed that can be charged by solar power and that is small enough and with a large enough electrical density that it can be used for powering the cooling units the firm produces and sells.

The Technology Transfer in our case is made in order to allow the firm to develop an improved product that will give it an advantage in the market compared to the competitions, creating a competitive advantage for the firm. The technology the university will transfer to the firm, the firm will need to further develop into an integrated product that it can later produce and sell.

In the case mentioned above the transfer is deemed to be serial from a university or a knowledge provider to a firm, an entity active in the market (see Figure 1 - linear knowledge transfer). That is not the case when dealing with two firms active in the market cooperating in the development of a joint technology. In such a case, the knowledge transfer is clearly two ways (see Figure 2 - Dual Direction Knowledge Transfer), between the two firms and therefore requires both firm to cooperate both in divulging knowledge and in receiving it. While that may be clear and obvious, one needs to consider that the knowledge transfer between the duo of university and firm can also be two ways. That the university may need some of the knowledge the firm has as input for the development of the new knowledge, it may go beyond that.

Figure 1. Linear knowledge transfer

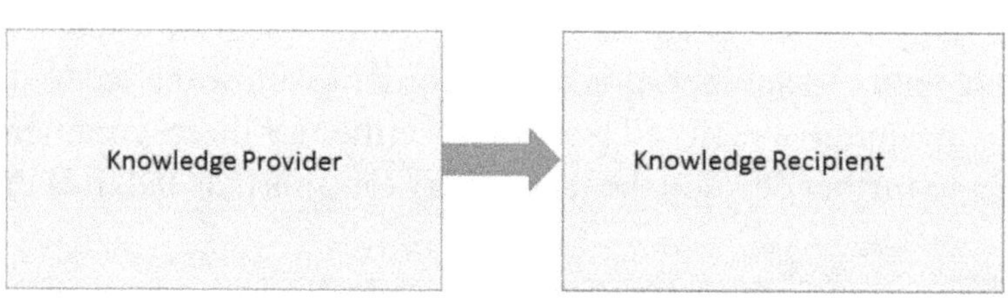

Figure 2. Dual direction knowledge transfer

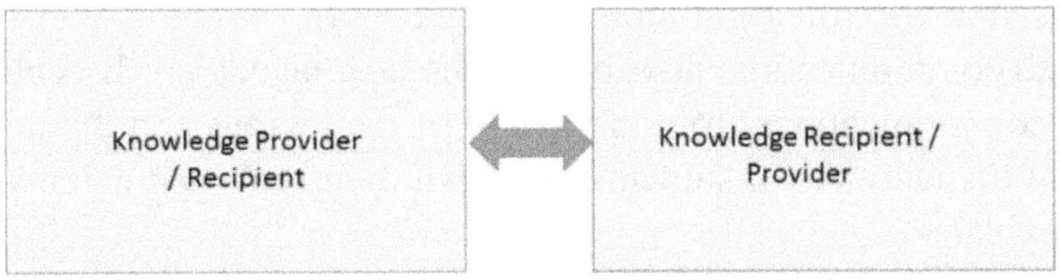

To sum the basic requirements for knowledge transfer, there should be at least one knowledge provider and one knowledge recipient. The participants may play more than one role, being both provider and recipient, or one may play two roles while the others play one or both. There cannot however be a participant which has none of these roles, except for the agency, which we consider as a viewer and not participant in any case.

ASPECTS

The most important part to remember is that Technology Transfer is not a value by itself but rather a tool, a means to an end. Therefore, the most important aspect of it is that the Technology Transferred must be usable. And with that aspect we shall deal below in the following subsection. However, there are some aspects of the knowledge transfer which we need to discuss at this point.

The best example to illustrate them is cooking. While in cooking there is tacit knowledge, the recipe which includes the list of ingredients and the steps required for preparation of the dish, the way in which the steps are executed is in many cases non-tacit (e.g. 'folding the yokes'), and therefore requires additional effort in order to perform the actual transfer, than just handing over a list.

Like in a cooking class therefore there is a need for some direct contact between the people studying in the class (the recipients) and the teacher (the provider). That direct contact requires a direct means of communication, a set of rules for that communication, the existence of the knowledge in a form in which the provider can use to transfer it, and trust.

If the knowledge is not in a transfer-ready format, that is in our case a combination of verbal explanation and demonstration, it would not be possible to hold the class. Imagine trying to do so without having a mutual language. The same applies if there are no tools for demonstration, it is very difficult to explain how to gently beat the eggs into a soft foam like substance without demonstration with the right tools (and the eggs).

Agreed communication, goes beyond the language, it consists also of the rules of communications from the length of the lessons and the number of breaks, to the fact that all students lay down their tools and listen when the teacher explains.

Last of course is the trust. It is important that the students trust the teacher that he reveals all they need to know and that all can trust the others to keep the rules of the communications.

These last aspects form a set of relational assets which can be acquired by participating in the process and learning from it.

It is therefore not surprising that organizations and units which have participated in such collaborations in the past tend to do so again. The experience creates two advantages for future participation in collaborations. The first is the removal of apprehension or any fears which the potential participants may feel. They have done it once and they know that it is not so terrible, and even know how to overcome some of the future obstacles. The second is the creation of the skills to form the rules for communication and to initiate the trust with other organizations.

In order to capitalize on such social assets, organizations tend to create units that specialize in being the interface between the host organization and the other organizations, and thus allowing collaboration via specialization. We shall deal later on with the role of the Technology Transfer Unit (TTU) and its different forms. One of the major attributes of the TTU, relevant to our discussion at this point is the specialization in liaising between the mother organization and its often-unique culture and the rest of the world. In that aspect, one would expect the TTU to develop preferences to projects and programs as well as partners with which it is already familiar, has relational assets and has accumulated the trust and experience.

In many cases, there is transfer of knowledge, mostly tacit knowledge from the firm to the university, including, standardization information, methodology and common practices or preferred solutions, but also preferred technologies and methodologies. In that aspect, it is important to recall that the industry has preference to specific methodologies or techniques when dealing with new products in order to fit the preferences of the market and regulatory demands, and these are normally not mandatory for academic research. However, in order for the results of the joint effort to be usable for the firms it is important that the academy utilizes accepted methodologies in its own work.

Technology Transfer: Example 2 - GLP

A university research group identified a cancer target and aims to develop it through potential molecules that will serve as drug candidates and will be tested in cells and in animal models. However, the finance of the research is best supplied by a program requiring the involvement of firms in the projects. The firms that are potential licensees of the results require in order to do so, to have results that are usable knowledge for the regulatory approval. They require that the university research group adopt Good Laboratory Practice (GLP). In order to be able to find firms to participate in a joint project the university adopts the GLP.

As mentioned above, the usability of the knowledge is a key aspect of the R&D projects, and will be further discussed below.

CHARACTERISTICS OF THE TECHNOLOGY TO BE TRANSFERRED

Here the discussion deals with the need for usability of the technology, the importance of transferring all relevant knowledge, the issue of making sure that the knowledge allows the return on investment and is protectable, the FTO of the knowledge, future developments, transferring the capabilities as well as the knowledge.

As mentioned above, the main aspect of technology Transfer is the usability of the knowledge transferred, however, this is a general statement. This point requires some further clarification.

It is important to recall that the process of knowledge transfer is costly. Both the provider and the recipient have to dedicate personnel and sometimes additional resources for the process, and the dedicated resources always have an alternative cost – people working on the R&D project, cannot be utilized to produce sellable products at the same time. Therefore, organizations are careful to invest their resources and efforts in transferring knowledge that has some significant importance, if not critical importance. While that can be obvious when discussing knowledge transfer from academic organizations to the industry, it is even more critical when dealing the knowledge transfer between two firms, active in the market. Such firms need to consider not only the aspect of resource utilization but also the exposure of their knowledge to potential competitors and the strategic significance this may have on their operations.

The case presented (see Technology Transfer Example 1) above demonstrates several characteristics required for the technology in order to make it transferrable:

1. The appropriateness of the technology to the firm. The need for the technology to be transferred stems from the need of the firms to keep or enhance their competitive advantage in the market, balanced against the cost of the knowledge transfer process as mentioned above. It is therefore important that the knowledge the firm expects to receive will be one that is required for its operation. The technology should fit the need of the recipient:
   a. It should fit the requirements regarding performance and standardization applicable to the firm and its intended product. That seems obvious, but it is not always the case. In the example above, the added battery should fit the relevant standards for batteries in such equipment, regarding energy consumption, safety etc.
   b. It should give the firm receiving the technology some advantage in the market that fits the market needs. If the market requires more energy for longer cooling periods, in the example above, than the technology maybe relevant, if however, it is not a real need than the technology will not create a real competitive advantage for the firm.
   c. The knowledge should be complete – that is to say that by receiving it the firm will be able to use it, and will not be required to do more research just to be able to use it. Of course, it may require some adaptations for the use, but the idea is that whole of the knowledge of producing the battery is transferred and not just part of it, in the example above.
2. The technology should be available for transfer – the technology provider (the university in our case) should have the knowledge free for transfer, with all the requirements for transferring both the tacit and the non-tacit aspects of the knowledge. This will be further discussed regarding IP below, but here we shall mention that the simplest case is when the technology provider owns the knowledge (e.g. by patenting it), and is free to license or sell it to the firm. There are other possibilities which are discussed in the IP section below.

In the example above we mentioned a case in which the technology to be transferred already existed in the technology provider and had to be validated and transferred to the recipient. However, there are cases when the knowledge is not yet developed.

Technology Transfer: Example 3

The same firm producing and selling the cooling equipment (see above Technology Transfer Example 1), identifies the need in the market for more cooling energy and decides to look for storage technology. It contacts a development center specializing in developing storage devices, and while defining a set of specifications contracts the development center to develop the storage device. The contracted firm starts work and within 6 months presents the cooling company with a battery according to the specifications it has received – fitting the solar charge demands, the cooling energy demands and interfaces according to the other operational and safety parameters defined. The delivery is of a prototype to be tested with the full production file of the battery developed.

The receiving firm tests the prototype and later incorporates it into one of its more advanced products for sale.

The above-mentioned example (Technology Transfer Example 3) presents a case in which the knowledge was developed specifically in answer to specifications derived from analyzing the market needs. As mentioned in the characteristics above, again we can see that the knowledge is fitting and complete, the developing firm has the right to develop and sell it (Freedom to Operate is assumed – more in the IP section below), and the receiving company has made sure during the definition of the specifications that it would answer market needs and therefore allow it to create a competitive advantage in the market. The main element that has changed so far is the fact that the knowledge to be transferred is also to be created first.

It may seem a trivial aspect but that is not so. There is always risk involved in new development. The risk is of course that the knowledge developed will not be usable and that the resources invested in the development, shall therefore, be lost. This can happen for several reasons which can be discussed later, but it is important to note that the more crucial the new knowledge is, the bigger the risk. The size of the investment required is also an issue but that is in many cases a risk that can be mitigated by stopping the project if it has not reached a certain milestone. The lost potential strategic advantage is dangerous and is more difficult to mitigate.

COMPLEX MODELS

The above presented examples were simple enough. They had two partners one the technology provider, and the other the recipient. The Technology Transfer was unidirectional from the provider to the receiver, and the provider had the rights to the knowledge, while the receiver had the access to the market. We have dealt with pre-developed knowledge and with technology developed on demand. Even the dual direction process was a simple case of two partners participating in the process on equal terms.

However, life is more complex normally. In R&D consortia where several firms participate, each is both a technology provider and recipient, making it important to assure it has the right to transfer the technology it provides and has the market access to utilize the technology it receives. In addition, there may also be research organizations that are normally just technology providers, but can provide the technology to several firms at the same time, for either competing products or for complementarity purposes. The research organizations would in most cases, especially when academic, not have a direct market connection and thus not be equal in their risk and potential benefit to the firms. Each additional partner at whatever role, or complex relationship or in-equality complicate the process and the interrelations among the partners. If, like in the EU Horizon 2020 (see website above), the partners are also of variable size, and from different countries (different cultures, legal systems etc.) the complexity raises even further.

For the process to succeed under such complex conditions and requirements, knowledge management capabilities and tracking become very important, as are clear cut rules, preferably simple, regarding ownership and usage rights of knowledge.

In such cases, clarity and transparency of the regulations regarding the management of the project and the resulting knowledge before the execution of the project, are not just demands of good governance, but basic requirements without which the execution of the project becomes very risky to all participants. It is important to enter a small remark at this point regarding the size of the participating firms.

While small firms may fear that larger firms will be able to usurp the knowledge of the small firms, take it over and own it to their benefit, the larger firms fear that the small firms may take the same knowledge and due to their swiftness and rapid decision making bring it to the market before the larger firms, taking over market share and sales. So, while both fears are legitimate, it would seem that the risk is real for both types and therefore both would require clarity regarding the regulations, to safeguard their interests.

The strategic importance of the knowledge to be transferred increases the effect at this point. Since the knowledge to be received, transferred or both has to have some strategic significance in order to justify the resources invested, the possibility of all of this being for nothing, increases the risk for the firms. In other words, since the knowledge is important enough to justify the investment of resources, it is also important enough for the firms to fear any risks regarding their ability to use it.

TIME

Time serves as a parameter which is further complicating matters. Time is mentioned here regarding the creation of the knowledge.

Since in many cases the knowledge created in a project is not what we would call "stand alone" and in fact some programs demand that there be prior research preceding the application to the program, it is important that the past IP ("Background Knowledge") is available for transfer at least to the extent that it does not limit the ability to transfer the new knowledge developed in the supported project.

This complicates matters especially when such background knowledge has already been transferred to other entities in the past for various purposes. In a simple case where all the partners are equal, they all create some of the knowledge and transfer it to the others, and they all have access to the market and are active there, the issue is less complicated. However, when dealing with a group of organizations, that are not equal in their access to the market, or in their ability to exploit the knowledge, but who are creating the knowledge, issue becomes more confused.

Research organizations, base some of their income on the "sale" of knowledge. When new knowledge is developed, which for commercialization to other entities require some rights in old, already transferred knowledge, this creates a problem.

This leads to one of the key issues in knowledge management in our context, and that is the tracking of the origins of the knowledge transferred from one entity to the other. Often strict regulations regarding access to the new and background knowledge apply, however, both recipient and provider had better be able to track their knowledge and identify deals concerning it in order to adhere to the by-laws of programs. In some sectors, human memory, with or without computer aid can be relied upon, but not in sectors with a high turn-over in HR. In such sectors, after a relatively short period there is no one left in the firm who can tell which part of the knowledge came from the outside, and which did not. In such case, advanced knowledge management systems are required, to keep track and maintain the ownership and usage rights of different participants.

SPECIAL CASE: THE TRADITIONAL SECTOR

The traditional sector needs some defining, when that term is used in this book the referral is to the sectors of the industry where R&D is rarely or even, not at all performed. These are not necessarily sectors with low technology. Quite the opposite in many cases. These are sectors where the innovation comes from equipment suppliers, from design or other sources, rarely from R&D performed by the production firms in the sector.

One such sector, is plastics. In this sector, many of the firms own a state-of the art manufacturing systems and equipment, their personnel are highly trained and proficient, but their innovation does not come from their own R&D and many do not have their own internal R&D departments. Of course, at the same time, in this same sector of plastics, there are other firms that do have extensive R&D departments and research new materials and their use and other possibility, but these are rather the exception than the rule.

Another such sector would be the jewelry sector, were again many of the firms, and most of the SMEs in that sector do not have an R&D department or R&D activities and they do not seem to consider them as mandatory, since the innovation in that sector, somewhat similarly to the plastics comes from innovation in design and catering to specific needs of target market segments, then R&D results.

Once again, there would be exceptions to the rule, firms that research new materials, coating technologies etc., in order to introduce innovation in additional ways.

The term "Traditional sectors" is used as these sectors contain many SMEs that are not start-up companies, but rather companies that have been around, probably family companies, for several generations, have ancient traditions and a presence in the market (e.g. shoes, furniture etc.).

The firms in these sectors are very good, else they would not have survived for so long, exhibiting efficient manufacturing, raw materials acquisition and in the marketing and sales of their products. They have their own way of doing things and in many cases, one cannot change the traditions in these sectors.

The firms in the sector as mentioned above would be very efficient in manufacturing and would from time to time perhaps, acquire new production equipment, but this is not their own development but something purchased from the suppliers. The suppliers of equipment and raw materials would perform R&D, and sell their products to the manufacturing firms, the ones with little or no R&D.

In order to explain the special aspects of working with such traditional sectors one needs to understand, that as long as the situation is as mentioned above, there are no problems, the problems appear when such firms, and even whole groups in the sector realize that they would like to introduce R&D into their everyday activities, and attempt to do so. Or when, from political, economic, and social reasons, the governing agency wishes to encourage them to take up R&D in order to increase their competitiveness.

Why is that a problem? Why not just establish an R&D department and start working?

The main reason is that establishing the R&D department by itself would not solve the problem, but could in fact cause an even more negative impression regarding the R&D activity in the specific firm, and thus reduce the long-term incorporation of such an activity into the firms in the sector.

Even though the establishment of an R&D department is not the only or even main limiting factor, since it is the most obvious one, it should be discussed. R&D requires personnel that is familiar with technological R&D in the relevant sectors. The problem with such sectors as described above is that there are not many people with the relevant research expertise. In such sectors as the demand for such personnel is relatively low, few if any aim to develop a career in that direction and therefore the supply of such personnel is low. There is, in most cases, some availability of research personnel from the academic organizations relevant for the topics of interest. However, such academically trained in research personnel is limited by the research organizational culture that trained it for academic research and not for industrial research and Development. When the number of organizations in the sector dealing with R&D is low, there is also a lack of people to re-train the academically trained R&D personnel for industrial R&D. Therefore, a major hurdle would be to find the right personnel and train or re-train it for industrial R&D activity that fit the sector. The more multidisciplinary the research in the sector the bigger the challenge as the larger the personnel required for the R&D department.

Another problem, derived from the above is of course, who will form the R&D department in such companies? It is important to understand that the firms in these sectors lack some of the R&D relevant management capabilities. As mentioned above, the firms are very good in procurement, production and marketing, each has found its market and knows how to maintain it. Therefore, these firms do not lack managerial capabilities in general, and are not necessarily low-tech. However, the do lack managerial R&D related capabilities. The firms have never before had to confront the issue of managing an R&D project on a full scale and in most cases, have no idea how to do so. Further to that, these firms have no idea how to manage the process of transferring the knowledge resulting from the R&D activity into execution in the production, or maintenance of the equipment.

It is not a trivial issue. When management does not know, or is not sure that it has the ability to manage the R&D process, it would try and avoid it, especially when it is used to do all other things very efficiently. One hears the following excuses: "it is too complicated for us"; "this is for other firms"; "this is very important and of course we are interested but now is not the right time;" and many other similar excuses, while the take up, or the proposal rate does not go up.

The problem as we described above start of course with having no idea how to form the R&D department, how to allocate resources set goals and monitor them and how to plan the interfaces with the other activities in the firm. It may seem like there could be a simple solution, to hire someone with the experience in forming or at least in managing such an activity, however here we run into the trap mentioned above – it is difficult to find such personnel in a sector with no R&D (Porath, 2012a; 2012b).

There is another problem, and that relates to the character of firms in such sectors, especially firms that are family owned, or that have efficiency based traditions. These firms do not relish the introduction of additional management members to manage an activity that is not clear to the rest. It would therefore be necessary to train the existing management and install in it the required capabilities.

To sum the above, when proposing a program for the encouragement of R&D activity in traditional sectors, per the definition above, it is important to recall the following:

- Forming an R&D department for firms that do not have one, will be required;
- The formation of such a department will require to search and acquire the right personnel and when necessary to retain it in relevance to industrial demands in the sector;
- Training the management in the managerial capabilities required for forming the R&D department and creating its liaisons with the other activities in the firm, managing the R&D results and forwarding them to execution in the production and /or maintenance activities in the firm;
- Setting the goals and process within the firm may take some time, and may require some external assistance. In addition to any R&D assistance that will be required in executing the work itself.

As mentioned above, in most cases there will be some relevant academic research on-going, and it is possible to try and recruit the academy to assist in the above-mentioned process of setting the R&D activity in firms in traditional sectors, however that is not so easy. It is important to recall that the academy can instruct in managing an academic type of research, and would have great difficulty in training in industrial R&D. Of course, in the setting of the interfaces with production and maintenance the academy would have even less to contribute, except theoretically in the management departments, which would be disconnected from the technical departments and faculties. It is therefore advisable to include some of the required assistance in the program in addition to funding. A component of consulting assistance in the formation and training of the R&D department, combined with some specific training for the management, would be advisable. Even with all these additions, the process would be more difficult than one would get with high-tech firms. Therefore, some preliminary or marketing activity should be incorporated into the program. The funding alone would in most cases not suffice.

REFERENCES

Porath, A. (2012a). Open Innovation for non-high-tech SMEs: The case of the User Association of Advanced Technologies program. In Rahman, H., & Ramos, I. (Eds.), Cases on SMEs and Open Innovation: Applications and Investigations.

Porath, A. (2012b). Theoretical model for a local economy open innovation program: A user organization. In *H. Rahman, & I. Ramos (Eds.), SMEs and Open Innovation*. doi:10.4018/978-1-61350-519-9.ch007

Chapter 3
The Unequal Partnership

ABSTRACT

In this chapter, we will discuss the unequal partnership between industry and academy when dealing with IP and its usage. The main parameters towards economic growth when dealing with knowledge creation and market introduction of said knowledge relate, to knowledge management and protection are mentioned. These issues refer to the ability of the academy to research all, but often the permission to commercialize only part (FTO), and the ability of the industry to commercialize combined with its limitations on research (Porath, 2010). Among other topics to be discussed are the knowledge gap between basic research and applied research, and the role of translational research. The role of basic research institutes, the role of industrial research organization (regional or national), and the competition between such institutes and the alternative, the R&D support program.

INTRODUCTION

The key for a successful program is to structure it in a way that will allow projects to be proposed, evaluated and executed, resulting in knowledge that can be taken forward towards the market. One of the key elements to constructing such a program is understanding the point of view and interests of each type of partner.

The misleading characteristic of the R&D activity is the usage of similar terms to mean something different by each of the potential partners. It is for that purpose that in this chapter the different types are described not only

DOI: 10.4018/978-1-5225-3652-9.ch003

in relation to each other but also as standalone organizations. At the end a special type appears, it is the individual entrepreneur which deserves special attention, although it is not the main topic of this book.

THE BASIS FOR THE INEQUALITY

There are many possible types of R&D support from the government or program funding point of view. These are mentioned elsewhere in this book when dealing with the different types of programs. However, there are some basic understandings common to them all.

While knowledge may originate from different sources, firms, academic research institutes, national and regional applied research institutes, to mention just the main sources, the tools to translate that knowledge into economic gains (including employment), are mostly just the firms. The firms being economic organizations, oriented towards profit making, and value creation are the natural translators of knowledge into economic and social tools, and therefore at the end of the R&D effort, will play a major role.

The same natural role ascribed to the firms, also includes certain organizational aspects which set them apart from the other sources of knowledge mentioned above. Therefore, whenever dealing with a combination of players in the R&D program projects, a combination of different types of organizations, the different organizational cultural aspects tend to present a potential conflict. While if the players all come from a certain specific single type of organization a conflict may arise from competition between the players. Of course, when dealing with a single player project there are no conflicts. Internal conflict between different departments in a single player, rarely come to the attention of the program.

Considering the above when designing or executing an R&D program the following potential conflicts (except for the single player programs) need to be considered:

1. Competition conflict – between players from the same type, easily evident among firms in the project.
2. Organizational culture conflict – between players of different types, the easiest to demonstrate is between academic institutes and industrial firms.

When dealing with a program supporting large consortia of both firms and different types of research organizations, both types of conflicts may exist and need to be dealt with in order to allow for a smooth operation of the supported projects.

ORIGINS OF THE CONFLICT: SINGLE TYPE

Whenever dealing with same type of organizations there is a danger of a conflict derived from competition between the players. It is easier to understand that a project in which several firms are involved, developing a joint technology that the firms are operating under the "fear" imagined or real of each other. The firms if involved in a specific project would have some link or affinity to the relevant sector / market intended to the technology to be developed. That is to say that they are probably to some level competing with each other in the market place. Even if to a certain level, they are not competing but are rather cooperating or even supplying each other, there is always the danger that they may become competitors due to the shared knowledge and similar abilities developed within the project. Another potential danger is that the larger firms may decide that the want to take over the smaller firms, once they identify their weaknesses during the cooperation in the project.

So, competition, or potential competition as well as taking over, whether real or imagined, are dangers that the program designer and executioner need to consider, and prepare to deal with, if they want the program to support functioning projects.

It is mentioned that the danger may be real or imagined as the firms in many cases have no predesigned strategies against the other participants, but have fears grounded in stories heard, or in past negative experiences[1].

It is important to note that these fears are normal at the proposal preparation and early execution phases, but that in most cases as the project continues and relational assets are created, trust develops and these fears subside. Therefore, the main dangers of these fears are:

1. Due to the fear from each other firms would not enter the consortium and the take up of the program will be so small, that the program becomes useless.
2. The firms will abstain from introducing into the "mutual pot" of the consortium real relevant technologies, at the proposal preparation phase, to the effect of rendering the project useless.

3. The fear will result in reduced real cooperation and willingness to share results in the execution phase, which will lead to the dissolution of the consortium.

The important aspect to bear in mind is that the creation of relational assets, learning to work together and to trust each other, both on the level of the firms, and on the individual level of the representatives involved in the project, are the best "balms" to reduce the negative affect of the fears. It is therefore not surprising that firms with positive experience in joint consortia tend to re-participate in future proposals, as they know how to overcome the fear, and how to benefit from the cooperation. In fact, the participation of such veterans may help in the development of future joint proposals.

The cooperation of academic research organizations has been going on for years (Kneller, Mongeon, Cope, Garner, and Ternouth, 2014; Hasan and Singh, 2016; Poo, Sander, Fivaz and Hayashi, 2008), exchange of researchers, and the demand in many places to perform a post-doctorate in another institute prior to entry into employment in a research organization, all contribute to a greater unity in culture among the academic institutes. Most of the fears mentioned above and potential conflicts, do not exist therefore in that type of single type cooperation.

It is in the best interest of the projects and of the program as a whole, to develop mechanisms that help the creation of such relational assets as soon as possible, to assure the smooth operation of the projects. Increased contact between the representatives, continued reporting of results, and consultations regarding the next steps, would help in fostering these relationships and in the creation of such assets. While keeping apart would in most cases result in slower development of relational assets and cooperation capabilities.

ORIGINS OF THE CONFLICT: MULTIPLE PARTNERS

Imagine the following scenario. A monastery is trying to cooperate in a joint community project with a bank. However, the values of the two partners, the training and outlook of the people employed in them and their ultimate goal set them apart. The amount of work required to cause two such organizations to work together for a joint project is big. In addition, while the monastery is operating within the parameters of its charter so to speak, the bank is probably being motivated by the wish to be better accepted by the community, which is only a side strategic goal for it. Therefore, the willingness to invest resources and effort is not equal between the two, their commitment to the project is different, and therefore what they would like to get out of it and eventually their satisfaction from the results would differ significantly.

It is clear that the monastery, with its high morals, social aim at a better more compassionate society, obedience to the superiors and piety is in stark contrast to the material gains aim, and profit seeking bank, where regulation and process are as strong regarding authority and where the social impact of action is secondary to economic gains.

Now substitute the bank[2] with another economic entity oriented towards profit – a firm; and the monastery with another more socially oriented organization – an academic research organization, and the potential conflict becomes clearer. That cooperation has been the focus of scientific research for some time, trying to understand it's parameters and the way to create fruitful cooperation (Porath, 2010; Jin, Wu and Chen, 2011; Shari, Haddow and Genoni, 2012; Kneller, Mongeon, Cope, garner and Ternouth, 2014; Parker and Kingori, 2016).

What we have just mentioned, is the difference in culture that results in a potential conflict as the different types approach the projects, with different intentions, and with a different definition of a successful project towards which they will strive and aim. The willingness to invest resources, and to make changes in the original organizational culture will also be derived from the same basic assumptions. However, the cultural aspect is not the only source of potential conflict.

When describing above the benefits firms derive from R&D, these were economic gains expressed in sales, revenues, profit, and market share. That is based on the ability of the firms to bring the R&D results to the market place, and capitalize on them.

When dealing with non-for profit legal entities, such as public universities, research centers and the like, the arguments regarding the derived benefits are empty. These organizations which for the convenience of the discussion we shall refer to as Research Organizations (ROs) are not interested in products' sales and market share in the technological sectors, further to that, due to their nature, they lack the abilities to bring the R&D results to the market. That is to say, that they do not have the production, sales and marketing, and post sales service capabilities, regarding products and processes derived from their R&D activities. They may have some such capabilities regarding teaching and perhaps regarding research services. They even may have departments or units titled similarly to those that exist in firms, but they are oriented towards teaching and research, not the direct market commercialization of products (Porath, 2010).

It is important to note that the Technology Transfer Unit (in whatever legal form it is organized) does not deal with the production and sales of products normally, but only of knowledge. Therefore, even that unit is not capable of bringing the results of the R&D activity to the market place by itself.

Thus, the first basis for the conflict is the cultural difference between the different organizational types, and the second is the inequality in the ability to derive direct economic benefits from the R&D project participation except the financial support granted through the program.

There are two important points to recall at this stage regarding the economic benefits that ROs derive from the collaboration. The first is, that ROs are economic organizations, even if they are not actively seeking profit. They are however limited in their activities by their resources and would always like to expand those resources. The second is that by employing the right regulations, the ROs can benefit economically from the transfer of the knowledge and rights for reimbursement. That is to say that different royalty mechanisms can be developed and utilized to help the ROs to derive economic gains from their activity.

Please note, this is not a claim that there are NO benefits the ROs derive from the projects, rather than their economic benefits are limited. The ROs benefit from the ability to enhance their research directly, their research capability (investment in the infrastructure and capabilities they operate) and enhanced prestige (depended to a large extent on their ability to publish). In many cases, they also benefit from the participation in the R&D projects by fulfilling the third column in the RO's roles, and that is the technology transfer or knowledge transfer to society (see http://www.astp-proton.eu/organisation/ for more information), which is considered evermore a requirement and justification for the public support given to these organizations. The difference in the ability to benefit, economically or otherwise from the cooperation is the potential basis for conflict, on top of the basic cultural differences.

However, in addition to the conflict inherent in the fact that both types do not benefit to the same extent from the cooperation, there are also the different point of view and goals, and additional cultural aspects to be specified below that contribute to the tension.

Confusion: Similar Terms Different Meaning

The title may seem confusing, but in that specific case it is much to the point. The terms used in firms such as head of department and chief of any kind, would normally denote the person in charge (responsibility and authority) of a certain activity. That person, e.g. head of department would normally be an experienced member of the department, and the leading or one of the leading experts in the field of activity of the department, and therefore the one directing its projects.

Such is not the case in the academic institutes. Due to the diversity of research interests in a department and the academic freedom to research topics that the researcher finds interesting, the heads of the departments rarely are experts in the fields of research of their department members. They are experts in their own research topics, but not of all their peers, and are therefore not directing or controlling the research projects these researchers are leading in their department. The term head of department denotes in the academic institutes an administrative rather than technological / scientific authority and a responsibility in many cases for the teaching activities, and much less so of the research activity. Due to the academic freedom in research the department head cannot direct the activities of his researchers or in fact researches that are performed outside his own lab.

In cases when the firms are not aware of such differences they may attempt to contact the head of the department and have him instruct his researchers as they would in other firms, however, that would not work in the academy.

Responsibility and Warrantee

In fact, even the term lab does not denote the same thing in both cases. The firm lab would contain projects that are of interest to the firm as well as relevance. There may be complete separation between the projects and the people working on them, and there may be safes with critical material stored in them.

The academic lab is a more open workplace where people may work together or assist others in the lab in the performance of specific tasks, in which they are more proficient. Lab meetings in which research projects are discussed are a normal thing, and rarely are research results kept in safes or out of the attention of the other lab members. Industrial or regional, non-academic research organizations may be more similar to firms, and more aware of the security aspects of industrial knowledge, but academic organizations are used to a more open method of research and instruction of research students.

However, the difference in control is not limited to the spirit of the work in the lab or to security issues, it relates also to the control the institute extends over the researcher.

The academic institute in the domain of research extends very little control or managerial control over the researcher heading a laboratory. It cannot compel the researcher to research a certain topic or to stop researching it, and further to that, due to the difference in expertise levels can control the work in the lab only to a very limited extent. Often, it can use only funding as directional tool for tenured researchers. That is very different from the control the management in a firm may extend over researchers in its R&D department and is important to understand.

Therefore, to demand from an academic institute responsibility regarding the work methodology and the quality of the work of a specific researcher, is in fact in many cases to demand from the academic institute to be responsible for activities it cannot control. The institute cannot control the way the head researcher is managing his lab, except through the budget, and that also means that the institute is very limited in the ability to control the quality of the work performed.

Warrantee

Another related important difference factor is the warrantee. Firms are used to receive warrantees regarding goods and services they purchase. However, as mentioned above, the academic institutes cannot supply that type of warrantee as they are not directing the work, and cannot compel the researchers in the institute to adhere to specific regulations and standards except academic ones, which are less relevant in most cases for the industry.

The researchers are allowed to research and test knowledge that may be under patent protection, as long as it is done for research and educational purposes, and without a commercial intent. That means that in some cases they may research topics that are protected and cannot be commercialized by anyone else, without being aware that it may be OK for academic work but not for commercialization, and someone else, not the researcher, would need to check the freedom to operate and to commercialize the knowledge. That cultural aspect, the freedom to research by curiosity is important to understand by firms, and to know that the knowledge coming from academy, is different in that aspect than the knowledge they may receive from other firms. There are units in the academy in charge of making such tests (FTO) but not the researcher. Often the FTO checking, entails costs additional to the research.

Publish or Protect

Researchers in the Academy live in a world where in order to advance their careers and to maintain it they need to publish articles in leading scientific journals. The need to publish stems among other sources from the structure of the academic institutes mentioned above. In most cases the peers of a specific researcher do not reside within the same institute, but rather in different institute in foreign countries. The peer evaluation system in which control of research results, and evaluation of the quality of the research are performed based on the publication of the work. It is so important that at some institute it is not allowed to approve a thesis that is not allowed to be published.

The need to publish is so strongly embedded, that even researchers who have reached the top of the academic career still publish regularly, and see it as an important activity even if it is clear that it would have little effect on their career.

In the industrial world, the firms do not view publication as a must. It can be a nice positioning tool, even a publicity tool, but it is in most cases not a requirement. Protection of the knowledge as either a secrete or as a patent are familiar requirements in the industrial world. As these two tools (trade secret and patent application) require delayed publication or no publication at all, this is an area for bitter arguments when trying to foster cooperation between industry and academy. If, as mentioned above, the university by-laws prohibit the approval of a thesis that cannot be published it means that the researcher cannot use his research students in projects where there is a ban on publication, and as these students are the main research HR – this effectively blocks the academic participation in such projects.

GENERAL GUIDELINES FOR DEALING WITH THE POTENTIAL TROUBLES

The potential troubles mentioned here are the culmination of the fears and apprehension the firms and the academic partners may feel when called upon to perform an R&D project, supported by a program. These needs to be dealt with as soon as possible, lest they may result in small take up of the program, a worse alternative, is that the take-up may end up with lawsuits and even lead to the cancellation for the program.

It is important to note again before discussing the tools to deal with the potential troubles that for firms there is in most cases the alternative of continuing their current activities. The longer the firms have survived up to the decision point to enter the project, the smaller is their tendency to prefer to enter the R&D activity.

One of the major points to illustrate this is the balance between accepting much needed finance, and the danger to the IP for Start-Up companies. Even in cases where the VC funding is low, and financing the start-up is a

Figure 1. The publication cycle

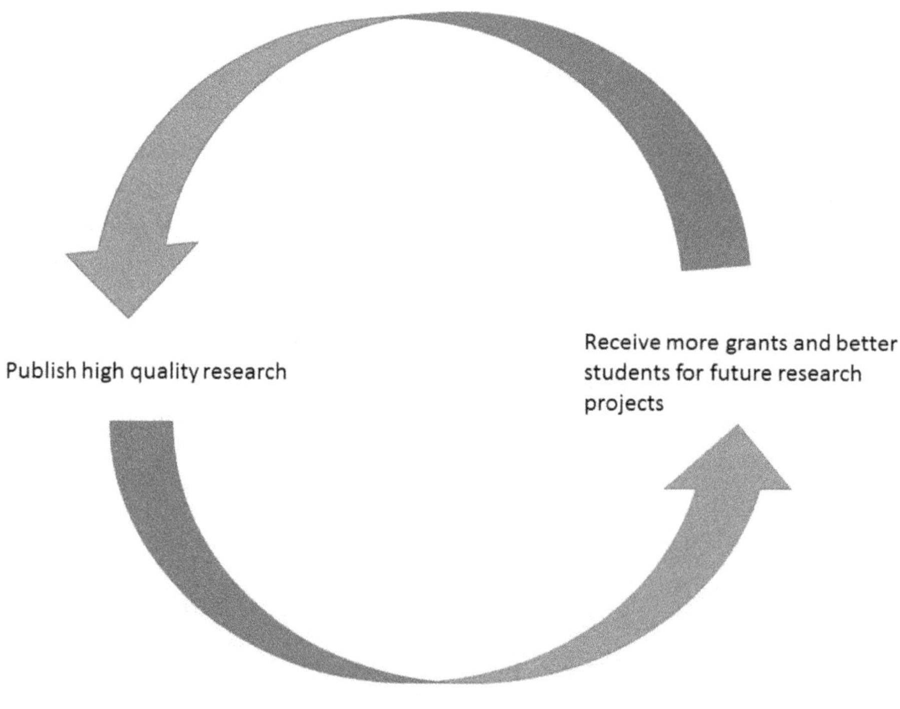

big problem, the entrepreneurs are often well aware that risking their IP is the worst thing they can do. The IP is normally the biggest if not the only asset and value creator a startup has, and risking that is the last thing the entrepreneurs will do. If they fail financially, but have the IP intact, they may still resurrect the firm, but if the IP is compromised, it is often the end of the adventure. Therefore, even in cases where the governing agency offers a substantial financing package, if the program involves a risk to the IP, future or existing, or even if there is a possible lack of clarity regarding the ownership and rights of the IP, start-ups will prefer to avoid the program altogether. This point needs emphasizing, any vagueness regarding the regulations or fate of existing or future IP, is considered a risk to the IP, and must be avoided.

Among the most common tools for attracting participation are:

Clear and Simple Project Guidelines

The requirements for simple and clear guidelines may seems self-explanatory, but deserves nonetheless some additional explanation. The clear guidelines, including the demand that they be published and be available before the proposal submittal, allows potential partners to better understand what they are getting into, which partners they still have to fear and to what extent. The resulting "fear management" allows potential partners to create their own red lines that define for them under which circumstances (e.g. other partners), they will continue with the submission, and when to opt out. Such guidelines and clarification will allow firms to manage their risks and enter or leave the consortium – making the project manageable. That is to say that a firm may declare that it will stay on, as long as a certain firm or type of firm does not enter, which will allow the rest of the consortium to develop entry and leaving conditions beforehand.

The guidelines could refer to different aspects, but would in most cases need to refer to:

1. Ownerships and right to use and to commercialize the resulting knowledge, and in many cases regulations regarding the right of use of prior knowledge;
2. Regulations regarding the management of the finance available from the program as well as the reporting mechanisms of the partners;
3. Management structure, and especially the definition of the relationships between the partners, the definition of the work plan and the knowledge and financial support;

4. Conflict resolution mechanisms. This should be very explicit and clear, making sure that it has the same relevance, as much as possible for the larger firms as well as the smaller firms;
5. The role of the governing entity in the management of the projects;
6. Offer or demand the written contract agreement between the partners, setting the basic rules for such agreement.

Another aspect regarding the need for clarity of the rules stemming from the inequality of the members is the haggling over the payments for the usage of the knowledge. As mentioned above the only normal way for the academy to see any financial gains from the R&D activity is by commercializing the knowledge in return for royalties (or equity). However, the academy faces a serious problem when discussing these royalties with the industrial members, as the Academy is not an active player in the market.

While it is sometimes difficult for firms, active in the market to assess the value of the ability to use the knowledge, it is much more difficult for the Academy, not involved in the market and with no intuition in the market, to assess the value for the knowledge. Industry often complains that it is difficult to deal with the academy as it is very demanding and the prices it sets are not realistic. That stems from the fact that often the only indication for the academy that the price demanded from the industry is too high, is the evident despair of the industry. This lack of familiarity not just with the market, but with the internal costs of the knowledge as well as the benefit for the industry, are obscure when it comes to the academy.

While this may seem very theoretical and can be regarded so at the beginning of a project, and even earlier at the proposal preparation stage, and therefore a bit unnecessary and useless argument, it must be noted that it is easier to agree the details between the partners at this stage than later on. Once both sides have invested efforts and resources, it is more difficult for them to agree on the prices and costs of the knowledge. Of course, both sides may feel that it is difficult to agree on the value of an unhatched chick, but at least an agreement about the process, and the setting methodology of the price should be agreed. This, the agreement of a process, a methodology, a mechanism of arbitration or similar is easier to reach at the proposal period or at any point before the work actually starts, than when research results start to appear.

One of the main functions of the governing agency would be to introduce the regulations compelling the partners to agree on these process and methodologies before the project starts.

Oversight by the Governing Entity of the Project Behavior of Partners

One of the most important aspects of the governing agency (Porath, 2010) is of course the mitigation of the risks involved in the R&D projects, by the financial support. For our discussion here, it is also as the supervisor and "parental oversight". Behavioral economics (Ariely, 2009), describes that it is human nature to better behave when observed by higher entities, but in our case, there is a stronger incentive. The governing entity can have a stronger influence over the good behavior of the players, as it could disseminate information regarding "bad" behavior, it could utilize economic tools to punish bad behavior such as finning the offenders, barring them from future participation in the program, and also other programs, and all the way up to legal action taken against them.

The above infers that the governing entity should have real powers over the participants. That "real" power can be the financing of current or future projects, can make use of legal tools and can be limited to the ability to state the events as they occurred. It also has the benefit of being there for the duration, and can recall past behavior when required.

In some cases, a combination of all of the above is possible.

Example 1: Oversight

In an R&D supportive program intended for the validation and technology transfer of knowledge from academy (single entity) to industry (single entity), the by-laws state that the academy appears as a sub-contractor of the industry and that the financial support goes through the industry to it. Therefore, the financial support is participatory for the industrial firm, but is 100% for the academy receiving full payment from the industrial participator. While the financing agency can exercise, the different powers mentioned above, it also insists that the academy and the industry sign an agreement – R&D subcontracting agreement. That is to allow the academy to use the legal tool (the contract with the industry), in order to claim payments not received but owed from the industry, in addition to any sanctions the agency can enforce.

The above-mentioned example is intended to give the academic partner additional recourse that will allow it to get paid – but will also allow the agency to avoid the need of preparing contingency funds to use for the academy's benefit in case of default payments on the side of the industrial partner.

While there may be little need to actually exercise such measures, just the fact that the governing agency is there and watching the projects, ready to intervene is a strong enough incentive. This role and the readiness to intervene with the relevant tools and measures, should be published with the guidelines before the proposal submittal, and can have a calming effect that will allow the participation of potential partners which may otherwise abstain from doing so.

Required Official Agreement of the Partners to Adhere to the Rules

The role of the official mutual agreement between the partners is to further support the other measures mentioned above. It creates a further legal bond among the partners while also setting in contract form the relationships between them and the regulations and by-laws for the internal governance of the project. By accepting these, the partners state that they accept the regulations and by-laws and that thus they accept the risks involved in following them.

It is important to note that the program must insist on the approval of such agreement before the execution of the project[3]. The approval should be subject to the condition that there are no by-laws or regulations included in the agreement that stand in contradiction to the by-laws and regulations of the program. This demand will assure that there are no critical demands included in the agreement by the more powerful partners that would have caused the other partners not to participate in the proposal if they were known earlier. In fact, if well published before the proposal submittal this will reduce the chances of even trying to pull over such a move.

There is an additional benefit from the point of view of protecting the partners from each other and that is creation of a legal obligation by the signing of such an agreement. If in addition to signing whatever agreement the partners have to sign with the funding agency, they need to sign one among themselves, it will give the partners legal recourse in case of problems without requiring the legal intervention of the governing agency. Each partner could use legal recurs for breach of contract according to the agreement between the partners. This is especially relevant to national programs. International legal action is a rare move, but the mere existence of the contract is a somewhat guarantee of adherence to the rules.

In order to reduce the chances of legal entanglements it is recommended to demand that several topics be addressed in the partners' agreement:

1. Project management structure – giving each partner a voice in the management of the project and a chance to protect its interests;
2. Setting of clear IP and publication by-laws in accordance with the by-laws of the program;
3. Financial support division and regulations regarding reporting of expanses – setting clear rules and governance structure;
4. By-laws regarding changes to work plan addressing content of the work, the budget and the timetable;
5. Conflict resolution mechanism that is fair to all partners and that allows the governing agency to intervene when necessary;
6. Special sensitive issues (e.g. "stealing" employee's Proton's website) that may danger the operation of the project;
7. Specific obligations to each other that may continue to be binding after the project has ended.

THE KNOWLEDGE GAP BETWEEN BASIC RESEARCH AND APPLIED RESEARCH

The Basic Research Aspect

The basic research is as a curiosity driven research an effort to understand the world about us, the ruling laws of nature and the actions of creature and natural processes. It is therefore an effort to make sense of the environment, the same effort that humanity has been doing since the dawn of time. It is the effort to make sense of the world around us without making use of gods and spirits, to replace meta-physics with physics.

The Role of Basic Research Organization

The effort of performing the basic research lies with the basic research organizations, such as the universities and the basic research organizations (e.g. Max Plank). They are organizations that aim to extend the boundaries of human knowledge and deal with research for research sake. They often comprise the best scientific minds that have been educated and trained to perform high level research, and to publish it, thus 'exporting' it to other researchers and humanity and also winning acclaim for the prowess in performing research. The universities as part of their role in society (Drew, 2011; Heinzl, Kor,

Orange and Kaufmann, 2013) are leading entities involved in this aspect and serve also as the primary source of future generation of such researchers.

The non-university like organizations, such as Pasteur and Max plank to name but two, are focused on the research, and may employ or supply research students with the opportunity to perform research, but cannot do so on their own, and need to partner with an academic title awarding entity (a university).

Applied Research Aspect

The applied research is an effort to show the feasibility of a certain concept or understanding resulting from the basic research. It is a preliminary step before the move towards development in which the final product will be designed and tested to show compliance with standards, regulations and market restrictions.

In some countries, where there are no liaising organizations (e.g. TNO in the Netherlands), the universities perform this aspect of the research as well as the basic research, being closet to the understanding of the principals resulting from the basic research, and owning the required infrastructure.

The Role of Industrial Research Institutes

The industrial research institute role, where they exist, is to perform the applied research and prepare the ground for the development to be done by the industry (Wolff, 1993; Elwood and Rainnie, 2012; Yang, and Jung, 2015). They need to be able liaise between the basic research performers and the industry in order to allow the knowledge that was created as a result of the basic research to be transferred to applications and products which will benefit the public.

This special expertise requires them to invest heavily in the equipment and specialized personnel of basic research, while at the same time maintaining the ability and contacts with the firms and the challenges the industry faces in the market place and with the regulation, and standardization in the relevant sectors. They need to develop their relational assets in both the basic research institutes and their academic culture and the industry with its own commercial culture.

The Role of Translational Research

In between the two, the curiosity driven attempt to make sense of the world and the specific application of that knowledge towards a product or a process, lies a small but deep gap. In this gap, some work is required in order to make the translation of the results of the basic research towards some application that can be tested. This specific step in the R&D process is called the translational research in which the results of basic research are translated into applied research which in turn will lead to product development.

The term "Translational Research" has developed in the biotechnology sector which due to the length of the R&D process and the different and significantly separated and defined stages of it, could present a clear and specific place for it in the process.

However, this term can apply, with a more or less clear definition to all sectors and technology, were there is a transfer from the basic research knowledge to an applicable technology. In some cases, it could be so short that the stage cannot really be separated from the other stages in the process (e.g. software, algorithm).

This stage requires an infrastructure for the research that is similar to that of basic research, but may also include a standardized equipment for production attempts that is more industrialized in nature.

An important note should be taken here, that certain knowledge and certain technologies may be applicable via several applications. The testing and proof of feasibility of one of them should be sufficient as a proof of feasibility for the entire technology, and following that each different application should be able to be developed directly towards a product.

The Industrial Research Institute's Place in the Chain

The non-basic-research national or regional research institute with the liaison role between the basic research and the industrial development is dealing in most cases with translational research.

For that purpose, it is important that it be able to communicate directly with the basic research organizations, have the relevant infra-structure and personnel to operate it, and at the same time be able to envision, plan and test different applications based on the knowledge derived from basic research demonstrating the feasibility proof of the technology.

For the purpose of our discussion here a proof of concept and a proof of feasibility of a technology are the same thing, and the nuances between them are not relevant.

When discussing therefore the research institute in the context of this section, we are referring to an organization that has developed a specific ability and as that ability requires a normally heavy investment in infrastructure, personnel and time, that ability is in most cases oriented towards a specific sector or technology.

The Dutch Polymer Institute is a good example (see http://www.polymers.nl/) regarding the ability of such a focused organization to promote not only new developments in the industry, but the actual transfer of basic research results to the industry. That ability to focus on a specific sector allows the institute not only to maximize the investment in the infrastructure, but also the effort and time invested in the creation of relational assets within the relevant industry.

We must accept that the relational assets both with the basic research performers and the industry knowledge receivers are important. Furthermore, that for the creation of such relational assets the organization should be able to identify with the cultural characteristics of each, and to be aware of the industrial needs and challenges while at the same time keeping up with the advent in basic research. That is not an easy task, and being able to focus on one specific sector, identifying and creating the relational assets with the relevant basic research organizations, and more specific research groups within these organization is a big help. As is the ability to link up with the relevant industry, preferably directly with the firms, but sometime also via association of the firms in the relevant sector.

Other organizations, like TNO and Max Planck would serve the needs of many if not all technological sectors. As can be imagined, this is a huge effort and requires a large infrastructural investment, not to mention the time required to establish the trust and relational assets required for the efficient operation of such an institute.

As shown by Max Planck institutes, the geographical spread of them over the region they are supposed to serve, in this case primarily Germany, is accompanied by a sectorial or technological specialization (Max Planck-Gesellschaft, n. d.). That phenomena results both from the effort to concentrate and conserve resource, by minimizing duplication of infrastructure and abilities, as well as from the need to establish one central point of relational assets focus for each technology. The industry should know where to go, and it had better be a specific place with the right contacts to the basic research organizations.

As we have just mentioned it, this is an advantage for such an organization, that while it is able to identify and stay in touch with the basic research groups creating the new basic knowledge, the industry knows how to find it, and contract its services, even new firms, or new comers to the region or the sector. This leads to a discussion of the main benefit of such an organization, which is the ability to efficiently contact it. This is but a single aspect in the general efficiency of the service.

The investment in infrastructure and the personnel required to operate it, as well as the investment in time and effort in the creation of relational assets lead to an increased efficiency and efficacy of the organization if fulfilling its role. The organization knows not only who to contact to get the information, and to whom to supply the results of its own efforts, but also acquires an expertise in the specific requirements of the translational research, in identifying and testing application for a proof of concept for a specific technology, how to transfer the research results in a way that is usable for the industry, and to include in the translation the industry specific requirements regarding standardization and regulation to a reasonable level.

In this aspect, the institute is a supplement to the R&D support program that is looking to sponsor translational research, as it fulfills its role and delivers the end product. There are however sever points to consider when discussing this point.

1. The institute shows increased efficiency regarding the translational research projects as it has the required infrastructure and personnel and the required relational assets to transfer the results to the industry;
2. It has no cultural problems internally or with the natural partners where required as it has overcome them in most cases over time;
3. It requires a heavy investment in infrastructure, personnel and time to make it efficient and is therefore a long-term investment or activity;
4. It is much more aware of the needs of the sector it serves as it is a long-term player in it and is especially familiar with long term major problems or challenges;
5. It is much less flexible due to the investment and relational assets than a national program issuing calls for projects, which may change its focus at will and still be relevant;

6. It has its own inertia as these are normally public organizations and cannot be disposed of at will, nor can their mission be changed at will, furthermore, they tend to develop the culture and ethics of public organizations, regarding budgeting and response time;
7. Due to its focus, it is much less apt to introduce solutions from external sector or technology into the sector it is serving due to lack of relational assets and in some cases lack of relevant research infrastructure. A program would more easily allow multi-sector players to cooperate in finding solutions;
8. Following the above, it is much less apt to offer multi-disciplinary solutions than a program, and may, in order to be able to do so, participate as a partner in a program.

The last point comes to illustrate that it is not necessarily a situation of either such an institute or such a program, but the two can co-exist. In many cases the institute would be there from historical reasons and the program would serve either as national tool relevant among other to the sector in which the institute operates, or as a boost to a specific sector. As can be seen from the experience in the EU Framework programs, such organizations tend to find R&D consortia interesting as they can cooperate with their natural partners under a program which seems to fit the abilities very well, and in which they are the best adapted partner.

THE INDIVIDUAL ENTREPRENEUR

The discussion above focused on the different organizations that are involved in the R&D process from the basic research and up to the final product development. One entity had been so far over-looked, the individual entrepreneur.

First let us define what not the topic of the discussion is. A researcher in a research organization that develops knowledge which is later commercialized via a spin-off with the research organization's assistance – is not the topic of discussion. However, if the research organization withdraws from the commercialization effort and at a certain point releases the knowledge to the researcher so he becomes the owner of the knowledge and needs to commercialize it himself. This last case, as well as a more difficult case when the entrepreneur develops all the prior knowledge or acquires it himself are the main discussion here.

The crucial characteristics therefore are the following:

1. The entrepreneur is the owner of the idea and of the knowledge behind it;
2. The knowledge requires further assistance in R&D towards commercialization to the market;
3. There is no firm yet, and probably some assistance will be required to support the transfer from an individual to a commercial legal entity;
4. The uncertainties are both technical and market oriented and therefore the program aiming to support that stage would need to be more comprehensive in the content it offers.

The situation discussed here is different from the others mentioned before due to the characteristics mentioned above.

The program to support the innovation in this case is special as it needs to support not only the technical development but also the setting up of the organization to bring the innovation to the market.

The project, as a whole would suffer from an increased uncertainty due to the extra activities required. In addition to the risks involved in the R&D itself, and the risks related to the market, which all other projects would bear as well. The setting up would involve legal acts such as a founder's agreement, the setting up of the company, which will involve setting up for example: files at the tax authorities and other legal and administrative actions specific for different countries and legal systems.

In addition to that, funds would be required, new partners would enter the group as part of the setting up of the company, processes would have to be developed and people hired. Each of these steps has the ability to slow down the progress, or even at the worst case to cause failure of the whole project.

Therefore, programs aiming at helping the individual entrepreneur should be able to support the R&D process, however, they should also be able to support the setting up of the legal entity to carry out the project, and support this action and direct it towards the best estimated solution. It should support the setting up of a firm, the processes of its operation and the hiring of its new personnel. This is important especially when the formation of firms is still complex. The program should also set some assistive tools for the fund raising required for the formation and initial operation of the new, as yet unformed, firm.

To top the challenges of such a program, it would also need to be able to evaluate projects proposals by individuals that will be carried out by the entities that have not yet been formed. It is best to prepare some guidelines for the firm formation, and the link to the project applicants.

Such programs do exist. A Technology Incubator program has been mentioned above. Other programs give partial support such as the SME Instrument, which will allow access to funds for proof of concept, and also assist in the access to VC. As are many programs aimed at assisting entrepreneurs in setting ups their companies. Most of the programs aimed at access to VC are intended for business concepts, and not strictly R&D support programs but some allow both, and most allow for the application to an R&D program either in parallel or as a follow up step.

In order to emphasize the special nature of the individual entrepreneur case, please consider the technological incubator program mentioned here. The program supports a group of preselected incubators, that house, and incubate projects that are later formed as firms. The incubator, according to the ruling model, is set up as a firm, and in turn, pre-selects the projects. The selected projects are submitted by the incubator to the program for support. The bulk of financial support for the R&D projects comes from the program, the rest from other private sources such as: from the proposing incubator, private investors including the entrepreneur or entrepreneurs.

Once pre-selected, and proposed for support and approved the project starts work, and is assisted by the incubator in the setting up of the company, in the administration of its activities and support of the reporting to the program as well as in the setting up of the accounting and HR. A main feature in the support is that the incubator serves as the administrative and financial home for the new firm, removing most of the formation and operation risks of the firm for the period it is housed inside, and prepares it for its activities after the incubation period. It allows the entrepreneurs to focus on their strong suite, the technological and business development of their new firm, and in some cases even assists in the development of the business side.

All of the above, does not go without a cost. The program is supported as a loan in case the firm succeeds, but as a grant in case of failure, therefore the royalties to the program have to be paid only if the firms succeeds to have incomes from sales. It also has a more immediate cost, that some of the ownership of the new firm goes to the incubator and supplement investors. In many cases this is the heaviest cost of the program, and at least one of the main reasons that some entrepreneurs are reluctant to enter the program, not willing to give up a part of the ownership of the project or firm from the start.

The incubator also serves as a consulting entity to the firms involved, allowing the entrepreneurs to benefit from the experience of the incubator firm and its managers, to learn some of the required management skills that they will need to continue and manage the new firm, and also puts them in touch with legal, accounting, patent and other service providers, that will later on continue to work with the firm.

The major advantage such a program has is the ability to support even first-time entrepreneurs, or entrepreneurs with little skills or an incomplete management capability set. The downside is the hurdle of the approval process and the cost it has in ownership of the newly formed firm.

It is perhaps advisable to have such a program complemented by another scheme that will support just the R&D of new firms, and perhaps give them some of the advisory benefits, but will be able to support more experienced entrepreneurs, requiring less administrative support and able to raise the funds, set the company and focus on their R&D.

As is mentioned several times in this book, there is no single magic solution or one single magic "killer application" program that fits all. A combination of programs is required in order to fully support the R&D process, each focusing either on a stage in the process, a sector with special needs and in several cases some overlap may exist – requiring coordination and control to make sure that no unwanted disadvantage is taken.

REFERENCES

Ariely, D. (2009). Our buggy moral code. Ted Talks. Retrieved May 2017 from https://www.ted.com/talks/dan_ariely_on_our_buggy_moral_code

Drew, S. A. W. (2011). Strategic roles of centers in business school development. *Journal of Management Development*, *30*(9), 793–815.

Elwood, L., & Rainnie, A. (2012). Strategic Planning in Irelands Institutes of Technology. *Higher Education Policy*, *25*(1), 107–129. doi:10.1057/hep.2011.25

Hasan, N., & Singh, M. (2016). Mapping the research collaboration of India and Vhina: A scientometric study. *Journal of Library and Information Science*, *6*(2), 207–218.

Heinzl, J., Kor, A. L., Orange, G., & Kaufmann, H. R. (2013). Technology transfer model for the Austrian higher education institutes. *The Journal of Technology Transfer*, *38*(5), 607–640. doi:10.1007/s10961-012-9258-7

Jin, J., Wu, S., & Chen, J. (2011). International university-industry collaboration to bridge R&D globalization and national innovation system in China. *Journal of Knowledge-Based Innovation in China*, *3*(1), 5–14. doi:10.1108/17561411111120837

Kneller, R., Mongeon, M., Cope, J., Garner, C., & Ternouth, P. (2014). Industry – University collaboration in Canada, Japan the UK and USA – with emphasis on publication freedom and managing the intellectual property lock-up problem. *PLoS ONE*, *9*(3), 1–19. doi:10.1371/journal.pone.0090302 PMID:24632805

Max Planck-Gesellschaft. (n. d.). Max Planck Institutes. Retrieved from https://www.mpg.de/institutes

Parker, M., & Kingori, P. (2016). Good and Bad research collaborations: Researchers views on science and ethics in global health research. *PLoS ONE*, *11*(10), 1–19. doi:10.1371/journal.pone.0163579 PMID:27737006

Poo, M., Sander, L., Fivaz, M., & Hayashi, Y. (2008). Asian promise: The state and future of collaboration in neuroscience. *Nature Reviews. Neuroscience*, *V9*(11), 881–884. doi:10.1038/nrn2512 PMID:18843270

Porath, A. (2012). Open Innovation for non-high-tech SMEs: The case of the User Association of Advanced Technologies program. In H. Rahman & I. Ramos (Eds.), Cases on SMEs and Open Innovation: Applications and Investigations.

Porath, A. (2012). Theoretical model for a local economy open innovation program: A user organization. In H. Rahman & I. Ramos (Eds.), SMEs and Open Innovation. doi:10.4018/978-1-61350-519-9.ch007

Shari, S., Haddow, G., & Genoni, P. (2012). Bibliometric and webometric methods for assessing research collaboration. *Literary Review*, *61*(8/9), 592–607.

Wolff, M.F. (1993). Long term research to become primary focus of revamped Japanese government institutes. *Research technology Management, 36*(6), 2-3.

Yang, H., & Jung, W. S. (2015). A strategic management approach for Korean public research institutes based on bibliometric investigation. *Quality & Quantity*, *49*(4), 1437–1464. doi:10.1007/s11135-014-0091-y

ADDITIONAL READING

Bienkowska, D., Larsen, K., & Sorlin, S. (2010). Public-Private innovation: Mediating roles and ICT niches of industrial research institutes. *Innovation: Management Policy & Practice*, *12*(2), 206–216. doi:10.5172/impp.12.2.206

Carpenter-Song, E., & Whitley, R. (2013). Behind the scenes of a research and training collaboration: Power, privilege and the hidden transcript of race. *Culture, Medicine and Psychiatry*, *37*(2), 288–306. doi:10.1007/s11013-013-9311-5 PMID:23580139

Dittman, D. R., Hawkes, M., Deokar, A. V., & Sarnikar, S. (2010). Improving virtual team collaboration outcomes through collaboration process structuring. *The Quarterly Review of Distance Education*, *11*(4), 195–209.

Duysburgh, P., Naessens, K., Konings, W., & Jacobs, A. (2012). Collaboration in multidisciplinary, distributed research organization: A case study. *Higher Education Policy*, *25*(3), 267–288. doi:10.1057/hep.2012.13

Gunawardena, S., Weber, R., & Agosto, D. E. (2010). Finding that special someone: Interdisciplinary collaboration in an academic context. *Journal of Education for Library and Information Science*, *51*(4), 210–221.

Hara, N., Solon, P., Kim, S.L., & Sonnenwald, D.H. (2003). An emerging view of scientific collaboration: Scientists' perspectives on collaboration and factors that impact collaboration. *Journal of the American Society for information Science and Technology, 54(10)*, 952-965.

Liu, B., Tang, N., & Zhu, X. (2008). Transferring technology across borders: institutetional effects in Chinese context. *Journal of Technology Transfer, 33*, 619-630.

Medicine, F. (2008). Critical issues for effective collaboration between academia and industry in the field of regenerative medicine in Japan. *Future Medicine*, *3*(4), 497–504. PMID:18588472

Osemene, K. P., Ilori, M. O., & Elujoba, A. (2013). Assessing the commercialization of herbal medicines research and development outputs in Nigeria. *Research Journal of Pharma and Technology*, *6*(6), 622–631.

Rosenblatt, M. (2013). How academia and the pharmaceutical industry can work together. *AnnalsATS*, *10*(1), 31–38. PMID:23509330

KEY TERMS AND DEFINITIONS

Academy: Any public or private, nonprofit research organization whether including education activities or pure research. Examples include universities, colleges, national or regional research organization and others. An important characteristic is that the organization is not producing and selling products derived from its research activities.

Consortium (pl. Consortia): A group of companies with or without academy organizations the has come together for a joint purpose, in our case here for the purpose of a joint R&D activity. The group may come together under an agreement, a contract or be incorporated as a legal entity.

Legal Environment: The legal infrastructure and the laws under which the R&D executors and the agency managing the R&D support program act. This includes among others the relevant incorporation laws, tax legislation, and Patent laws to name a few.

Technology Transfer Unit (TTU): A specific unit within an organization, normally a research non-profit organization, charged with the protection, and commercialization of the knowledge resulting from the research activities of the organization. In can be an internal unit within the organization, it can be an affiliate organization, or even an external organization to which this activity has been out-sourced. It is important to note that unlike patent advisors this unit is also charged with the commercialization of the knowledge and not just the protection of it. TTO – a Technology Transfer Organization, will denote the case of a separate legal entity.

ENDNOTES

1 The author participated in a kick-off meeting of an R&D consortium for R&D development meeting in which the large firm leading the consortium confessed that it had apprehension that the smaller firms, being quicker to the market may use the knowledge to beat the large firm to the market, and take over some of its market share. Following that the smallest firm in the group confessed, that it was worried lest the large firm would take it over. It created a sort of fear balance which was later replaced by trust as the project went on.

2 The substitution is for clarification purposes, it is possible to have a joint project between a bank and a monastery.

3 As these agreements may have a high cost in time and money it is not advisable to demand them as part of the proposal which may not be acceptable, but rather as a pre-execution condition between the approvals of the proposal and the project execution.

Chapter 4
Knowledge Packaging and Knowledge Protections

ABSTRACT

This chapter deals with some of the basics of knowledge protection. The different possibilities and the specific cases they fit. It deals with the different possibilities, the characteristics, the advantages and disadvantages each technique offers, and some of the current trends. Examples help illustrate some of the points argued. The purpose of the chapter is to familiarize the reader with the basics of knowledge protections and create the common grounds for further discussion.

INTRODUCTION

The process of commercialization, in lack of better term is the process in which the research results of an R&D project or activity, are brought to the market as products, process or services. While this simple explanation may seem clear enough, it is in fact a large family of different activities leading to a complex end. The simple view of taking the research results of a potential drug and turning them into a pharmaceutical approved drug is long and cumbersome (Rosenblatt, 2013) and may include in addition to the approval other aspects, such as the production process and compliance with GMP[1], or with other regulatory issues. It is a lengthy process and the product may fail at each stage, never reaching the desired end.

DOI: 10.4018/978-1-5225-3652-9.ch004

Another product in communication for example may follow a simpler path, it may need to show compliance with communication regulation demands which are simpler than FDA demands, but it too will have to undergo the move into production, adhere to quality standards and then enter a competitive market.

With processes and services, the path may be less complex, but just if not more difficult.

While all the work described above is ongoing additional knowledge to the basic knowledge of the potential drug or other product is created. This additional knowledge is also valuable, both regarding the time and effort invested in developing it, but also in certifying it.

Economic theory has long stated that there is a need to create an advantage in the market. The current and future basis for that advantage is the knowledge (Porath, 2010). Since knowledge is consumed as any resource, it is the ability to recreate it and to sustain the advantage that the new knowledge creates that is important.

KNOWLEDGE PROTECTION: THE TOOL FOR COMMERCIALIZATION

Since knowledge can create the advantage in the market, it is important that it cannot be stolen or taken by the competition. Knowledge that can be easily copied is of little strategic value as it gives only a short-term advantage in the market.

Therefore, knowledge must be defined and protected (Porath, 2010; Fuller and Hohman, 2010; Kundu, Bhar, and Pandurangan, 2015; Badea, Tanasecu, Marin, Stephanescu, Vladut, Bucur, Ciocanel and Ivan, 2015; Vinig, and Lips, 2015), and the protection can affect the ability to successfully utilize the knowledge and transfer it. Let us take a moment and look at our drug example from above. The approval process for a drug is lengthy and costly. To go through the entire process and then to find out that since the drug composition and manufacturing cannot be protected, it can be copied by the competition, means that no company would take upon itself the approval of such a drug candidate. A firm that would do such a thing would not be able to recuperate the development and approval costs quickly enough to avoid going under. Therefore, if you want your drug candidate to have any chance in the future of reaching the market, the knowledge related to it must be protected from the competition from the first time that there is any knowledge worth protecting.

So, the protection is there in order to allow the firms that take the knowledge and invest in turning it into a marketable product (process or service would be similar, so we shall focus on a product), by making sure that at least for a period the firm taking the knowledge would be exempt from competition and thus the new knowledge will allow it to produce enough economic gain to justify the cost and effort invested in the new product.

Figure 1 explains the process from the economic point of view. The investment required to protect the knowledge makes it worthwhile to proceed with the development of the product by the approval process (which may include additional new knowledge that needs to be protected as well to increase

Figure 1. Knowledge protection cycle

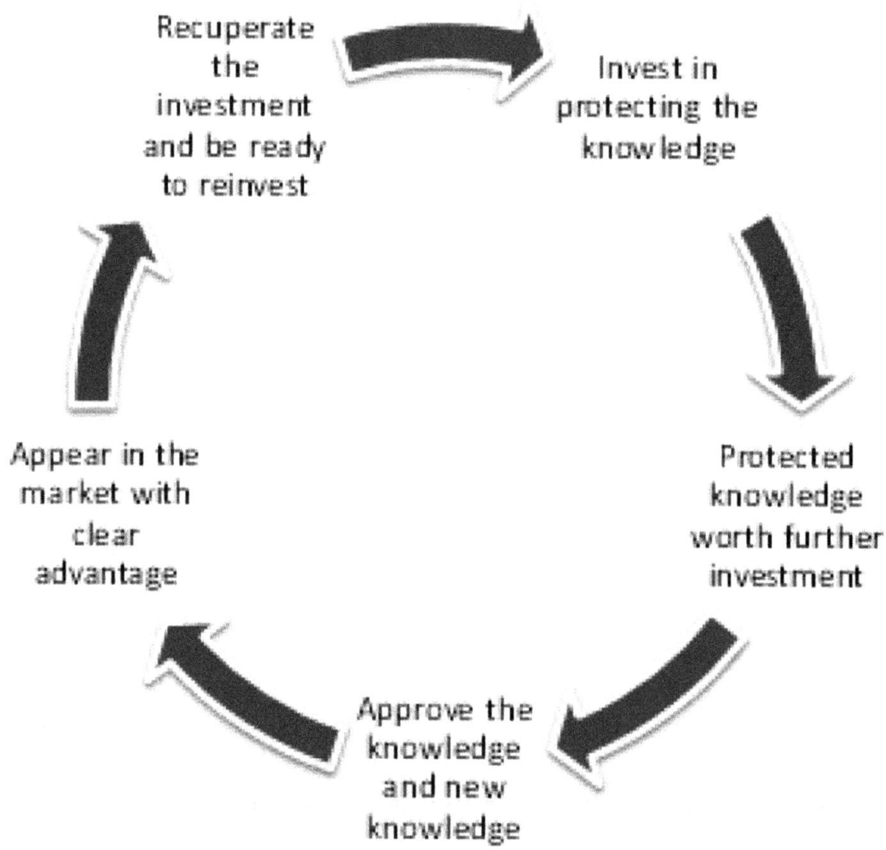

the value of the knowledge and to make copying it more difficult). Once the approval process is over and the product can be moved into the market where its advantage and the fact that it cannot be copied creates the opportunity to accrue economic gains which in turn will both justify the initial protection of the knowledge as well as the entire process.

Of course, the process is described in simplistic terms. The initial protection may be done by one entity, and the rest by other entities involved. However, the principal that the initial protection allows the rest of the process is crucial to understand. The cost of protection of the knowledge is normally transferred to the recipient of the knowledge towards the market. So even if the initial knowledge was developed by a university and later licensed to a firm, the license will include the cost of protection which the firm will have to reimburse the university for. Therefore, it can be said that the cost is born by the firm after all, and it shall be ready to bear it and the rest of the costs only if it is able to recuperate them at the market. The entire process is economically justifiable only if the knowledge is protected enough to allow it to create a strong enough advantage in the market so as to allow the recuperation of the costs incurred.

One must bear in mind that at each stage of the approval the intended product may fail. Such a failure (technologically, economically, and regulatory) means that the product will not reach the market and therefore the costs incurred so far in the process cannot be recuperated by it. That risk has to be incorporated into the future revenue stream and justify the investment all the same.

Before we continue it is important to stop for a moment and consider the risks involved. The reasons for failure mentioned require some amplification.

The technological failure is in fact a risk on two plains. The technological risk may be expressed as a risk of the product itself failing in its ability to perform, in the example of the drug mentioned before it may fail to be safe enough, or active enough. It can also mean a failure in the production process of the product. The production of the drug may require GMP and failure to achieve that may mean the failure of the product. In other sectors, it may mean that the production failed to comply with production standards, each relevant to its specific sector.

The economic risk is expressed in the potential incurred costs. The production and especially the quality control of the production may present costs that could influence the future price of the final product and make it unmarketable. Or these increased costs may make the return on the investment period lengthy beyond the protection of the knowledge, which would make the entire project economically not viable. In order to keep this simple, we do not consider here alternative investments. Thus, the economic failure can be deemed to extend to both the costs and future revenues.

The last but not least risk, is the regulatory / administrative risk. Mistakes in the procedures, administrative mistakes, that disqualify the products from final approval may happen and prevent the product from entering the market.

Firms commercially active in the market, and used to receiving knowledge from external sources, know the level of the risks of their sectors. They also know how to calculate the right rate of return, the return period etc., and decide about the future of a technology or product. For our discussion at this point the specific parameters are not relevant. However, in a qualitative way, it is clear that the stronger the protection of the knowledge, the lower the approval cost, and the shorter the time to the market for the product, the more attractive the knowledge is to be transferred.

Of these parameters, the approval cost and the time to market are beyond the control of the original knowledge developer. The only parameter which is within his control at least partially is the protection of the knowledge and its strength.

STRENGTH OF THE PROTECTION OF THE KNOWLEDGE

When dealing with knowledge protection one should bear in mind that there are normally several means of protection available around the world in different countries. In order to make the discussion most effective and generally relevant we shall focus on the most prevalent means of knowledge protection.

The differences between the different protection modes refer both to the strength each offers regarding the legal protection – prevention of copying he knowledge, and at the same time each is designed to fit a different aspect of protection better.

However, one should bear in mind that the protection can also vary within a specific protection mode, according to the quality of the specific protection application, the sector of the technology and the different trends or fashions at the specific time the protection is created.

There is one additional aspect to recall regarding the protection of the knowledge. In order to protect the knowledge, in most cases there is a need to specifically define the knowledge to be protected. In this is a patent, or copyright, it must be made clear exactly what is protected and what is everything else. While this may at first seem to be a problem and a tedious and onerous task, it is also a tool in aid of the commercialization of the knowledge. It is easier to commercialize a well define well with clear content than a hazy unclear agglomeration of knowledge. The contribution of the packaging of the knowledge on the commercialization is discussed elsewhere.

One additional point to remember is that the knowledge must be usable. We have mentioned this above when discussing the different points of view between research entities and industry when dealing with researching patented knowledge.

The industry performs R&D in order to gain an advantage over the competitors in the market (or in the case of start-ups to create value in the firm), and therefore the results of the R&D must be available for development into products.

It is therefore important to understand the difference between patentability and Freedom to Operate (FTO). When the knowledge is developed, it can be patentable if it shows enough innovation to be considered a large enough step from existing inventions – not trivial.

If the knowledge is new enough to be patentable but relies on knowledge belonging to someone else, in a way, in which you cannot use the new knowledge without the old, then while the new knowledge may be patentable, there is no real Freedom to operate and to exploit it without the old knowledge.

Example 1: FTO and Patentability

As happened before several times, a drug used for a specific indicator seems to have a side effect which may be more profitable. The drug has been patented by company X, the basic formula and some of its derivatives, the purification and the usage all have been patented. Therefore, company X is the sole entity allowed to manufacture and sell the new drug. In that case, it is possible under certain circumstances, if a new use is discovered, by university Y for example, to be able to patent the new use of the drug. However, university Y will have difficulty when trying to license the new patent as no firm would be able to manufacture the drug and sell it, without permission / license from firm X. At best, university Y could offer the new patent to company X to enhance its production and further its sales. University Y may be able to stop company X from selling the drug for the new indication, and company X from anyone else producing the drug for any other purpose than the original – definitely a lose-lose situation.

The knowledge that interests the industry is the knowledge that can be utilized, that is the knowledge that has no constraints as much as possible. One should be aware that it is not impossible to utilize knowledge that is owned by others – getting the license or sub-license is a viable solution in most cases, but it does introduce another element of uncertainty that did not exist before. The owner of the previous knowledge can refuse for any reason it may like, including commercial reasons, to license the knowledge to other entities, and thus can block at least partially new inventions.

However, prior knowledge may not always be such a problem. If prior knowledge had been protected and the protection has expired, for whatever reason, that knowledge can no longer be patented, but it is open for free use, now that it is no longer protected. Patents going forward based on this knowledge can be protected, and can be utilized. In the example above of the drug, if the patent on the molecule has been neglected, it is possible to patent same drug for a different indication and actually manufacture and sell it for the new indication. In such a case, the old knowledge is considered in the public domain and as such open for use.

The important principal here is to make sure that the regulations of the new program are clear in their purpose, that the knowledge produced in the projects under the program is usable for the participants, without being dependent on the good will of other patent owners external to the program or the proposers.

Example 2: Partners in Knowledge

Two universities cooperate in the development of a new electro-optic component. They patent the invention, and later on one of the universities decides to enter a project with a company for development of the component for a specific use. The company and the university apply for a patent for the new component. However, when the company seeks an exclusive license on the old patent in order to have the basic knowledge all to itself and block the competition from developing similar components based on the same technology, the first university, the co-owner of the old patent objects, and presents commercial demands which the company is not willing to comply with. The negotiations continue for a long period and the competitors in the meantime developed competing components based on other technologies and the company has lost it leading edge in the market. In order to prevent such future problems, the universities resolve in the future to agree on licensing leadership among themselves including the power to license without objection (given to the licensing leader), and agreement on the division of the resulting income.

SECRET (TRADE SECRET)

Keeping the knowledge secret is perhaps the oldest method of knowledge protection. There is the old story of the mirror manufacturers in Italy who kept the secret of mirror production a secret and all who tried to reveal it were murdered. While this method may be effective (e.g. the Coca-Cola formula) it has some disadvantages. While it may appear cheaper, it is not necessarily so, if one has to go to special measures to keep it a secret. A restaurant that keeps a sauce a secret, would pay not only in increased transportation costs of the sauce from the production site to the different branches, but will also have to limit the users. People suffering from allergies could not risk the sauce as long as they do not know what it contains, people who avoid different products due to religious reasons would also be prevented from the consumption of that found and may avoid the restaurant altogether. Once the restaurant wishes to reveal that the sauce is safe for the different types of clients, by disclosing what is contains, it will lose the secrecy of the sauce.

Trade secret has one additional advantage, the longevity of the protection is potentially limitless in time, and is not contingent on international agreements.

One additional major drawback is that as long as the secret is kept as such, anyone who stumbles across the solution, by independent research may move either to reveal it or to try and protect it for his own benefit, in which case the original owner is left with nothing.

Copyrights

Trade secret is effective only if the secret cannot be revealed from the end product. Of course, for works that are published, literature, art, shows and movies etc., this is not relevant as the knowledge is published. Copyrights prevent unapproved copying, but it has its own limits.

It has a limited life span (different under different legal systems) compliant with the lifetime of the authors (with extensions in some countries post mortem). Copyrights relate to the work as it is presented and not to the general idea behind it. The fine line between infringement of the copyrights and legitimate usage of the ideas contained in a published work is often the case for court discussions. Therefore, in some cases this is considered a weaker protection, depending on how much needs to be changed in order to make another piece legitimate, while fulfilling the same function. When dealing in art and literature that may seem complicated, but consider the case of protecting a piece of software under such a protection. How much do you need to change the software to make the new piece legitimate, while fulfilling the same function? It is clear that protecting the idea (algorithm) behind it, in addition to the copyright would be stronger.

Patent

Protecting the idea, the technology, is what patents are all about. Patents are a dual system in which in each country there is a system for applying for, testing and granting patents, while under international agreements of recognition such patents can be extended to other countries (subject to further examination, but reserving certain rights such as the submission dates and priority).

Patents have some advantages – they cost money and time to approve, which limits the potential competition. They award a period in which the patent recipient is awarded a virtual monopoly regarding the commercial use of his invention, and they can be extended all over the world.

However, at the same time they have some disadvantages. Funny enough the first is that they cost money and there requires some investment. They grant their protection only for a limited time, and when that protection expires, the invention is laid bare for all to copy. If we look at the case of the competition in revolver manufacturing between Colt and Smith& Wesson, once the Colt patents expired Smith& Wesson entered the market with full force breaking the former monopoly on revolvers[2]. Patents can also be revoked through the courts, in which case not only is the invention disclosed to the public but the protection is gone, and financial costs are incurred as well.

At this point, one should bear in mind that patents do not offer the same level of protection in all sectors. In pharmaceuticals, they are very strong, and in that market patent infringements are easy to detect and legal action taken are very decisive. In the field of software that is more complicated and protection is therefore less strong. However, even within the field of pharma some patents such as those protecting the structure and the manufacture of a certain drug for certain diseases are stronger than those protecting only the usage for specific purposes. Therefore, if you are the patent owner of s specific molecule and its use for cardiac disease treatment, you have a stronger protection than if you are only registering someone else's molecule or one that is in the public domain for a specific cardiac treatment.

The down side of patents is therefore that they entail publication of the protected knowledge, which in cases of long term use, such as drugs, it means that once the patent period expires your competitors will use your knowledge to offer competing products at reduced prices – the generic drugs sector. In that specific case that downside is unavoidable as the approval would require disclosure of the drug structure and much more about it.

In other cases, (e.g. computer technology) the technology may be obsolete before the patent period expires and the publication therefore is less critical in that regard, but it still may help the competition to find ways around the patent and still cause trouble.

However, with all the disadvantages mentioned above patenting is such a powerful knowledge protection tool, that they have become in some sectors a measure to the firm's owning them potential in R&D.

PUBLICATION AS A KNOWLEDGE MANAGEMENT TOOL

While in many cases, the industry views publication as the enemy of strategic knowledge commercialization, that is a narrow view of the possibilities publication holds for the strategic knowledge manager.

In order to better explain this point, let us consider the protection means mentioned above, copyrights of course are helped by publication but they do not protect the exploitation rights of technological inventions and are therefore relevant only in a limited level when dealing with software protection.

Regarding trade secrets – publication negates this means and therefore is viewed as counterproductive in this case.

Patenting – here publication prevents the issue of a patent as the knowledge contained by the publication is considered as existing in the public domain. Unless the patent application (in which ever form selected) is served before the publication the knowledge contained in the publication cannot be protected alone by a patent. However, this point has its strategic value. As in open source, the fact that some of the knowledge exists in the public domain and cannot be patented allows many users to potentially work on it and promote the future utilization. This point is better illustrated in the following example – a scientist discovers a cellular cycle that could be used for treating cancer.

He knows that several pharma companies are working on the same cycle and that they may try and patent some of the molecules that disrupt this cycle so as to prevent other firms form using them to develop drugs – leaving the full potential just for them.

Example 3: Publication as Knowledge Management

A scientist publishes an article outlining each of the molecules involved in the cycle and its specific role, and some of the companies claim that this publication has ruined their future ability to develop drugs and patent them, making this option obsolete. Other pharma seeing that the cycle is explained and the relevant molecules, start synthesizing derivatives of the molecules changing them enough to allow them to patent the drugs, and use them on the published cycle. Other firms basing their work on the chemical knowledge regarding the molecules patent other applications (use patents) for these molecules. The scientist argues that in that specific case, publication has opened the field for further development.

In other fields, the publication such as in the sector of open source allowed more users to develop products by reducing the development costs and allowing developers to learn and benefit from the knowledge of each other. In some cases, publication can prevent future patenting, as a strategic tool to direct research and development.

COMPETING TRENDS

Regarding the issue of knowledge and R&D support programs using public funds for the creation of that knowledge there are conflicting trends.

The first trend, strongly supported by the industry and very much representing the main policy demonstrated in this chapter, advocates the availability of the knowledge and the need to protect it to commercialize; the second trend following the spirit of open source claims that as public funds were used for the funding of research therefore the same public should be benefiting from the availability of the results.

The second point needs some further elaboration. The logic behind it is that public funds were used to create the knowledge, and that therefore it should be made available to the public, especially the R&D community, for exploitation for the general good. This point further states, that since even the publication of the research results is very closed and costly, it should be made more accessible to the R&D community and not just for a single recipient, who may even, from commercial considerations, try to hid and not use the knowledge. Which would then be detrimental to the public good.

Recent trends in the EU have tried to find a balance between the two, making academic knowledge more "open source like" and open for access by other research teams, except in places where its commercialization (and that in a very limited way), could be damaged. This is also with the intention of lowering the cost of research by lowering the cost of access to literature (Porath, 2013). So, the main idea is that research results can be recycled for future development or even research and that such access should be the general rule rather than the exception. In the US a similar trend is also visible (see http://www.arl.org/focus-areas/public-access-policies/federally-funded-research/2696-white-house-directive-on-public-access-to-federally-funded-research-and-data#.WXwwrojyvIU).

In some fields, of course this could have potentially dangerous results. Pharma, as a long and costly approval sector would require the ability to protect the knowledge in order to allow for the recuperation of the expanses. Gaming on the other hand could be more tolerant. Therefore, the regulations in that regard, will need to consider sectorial characteristics.

There is no magical solution and some of the ongoing trends can be view in the published EU documents (EU, 2009), and one should see where this process stops and which balance point is reached.

REFERENCES

Badea, D.M., Tanasecu, F.T., Marin, A., Stefanescu, G., Vladut, G.C., & Bucur, D., CioCanel, B., and Ivan, I. (2015). Capitalization of scientific knowledge and technology transfer between public research institutions and the economic environment. *The Romanian Review of Precision Mechanics, Optics and Mechatronics*, *48*, 31–37.

EU. (2009). Making public support for innovation in the EU more effective – Lessons learned from a public consultation for action at community level (working document). Commission Staff.

Fuller, J., & Hohman, C. (2010). An analysis of the techniques of technology transfer. *Journal of Applied Business Research*, *26*(6), 1–10. doi:10.19030/jabr.v26i6.323

Kundu, N., Bhar, C., and Pandurangan, V. (2015). Imaging technology transfer: An analysis of intrinsic factors. *South Asian journal of Management, 22(3)*. Pp. 69-95.

Porath, A. (2012). Open Innovation for non-high-tech SMEs: The case of the User Association of Advanced Technologies program. In H. Rahman & I. Ramos (Eds.), Cases on SMEs and Open Innovation: Applications and Investigations.

Porath, A. (2012). Theoretical model for a local economy open innovation program: A user organization. In H. Rahman & I. Ramos (Eds.), SMEs and Open Innovation. doi:10.4018/978-1-61350-519-9.ch007

Porath, A. (2013). Reuse of excess research data for new researches. In H. Rahman & I. Ramos (Eds.), Ethical Data Mining Applications for Socio-Economic Development. doi:10.4018/978-1-4666-4078-8.ch003

Porath (2010). Collaborative Research in Management: Inside out. SAGE Publications, UK.

Rosenblatt, M. (2013). How academia and the pharmaceutical industry can work together. *AnnalsATS*, *10*(1), 31–38. PMID:23509330

Vinig, T., & Lips, D. (2015). Measuring the performance of university technology transfer using meta data approach: The case of the Dutch universities. *The Journal of Technology Transfer*, *40*(6), 1034–1049. doi:10.1007/s10961-014-9389-0

ADDITIONAL READING

Ding, Z., & Ng, F. (2010). Knowledge sharing in architectural design institutes: A multiple -case study. *Construction Innovation*, *10*(3), 267–285. doi:10.1108/14714171011060079

Dogra, R., Garg, R., & Jatav, P. (2013). Technology transfer in pharmaceutical industry: Transfer of process from development to commercialization. *International Journal of Pharmaceutical Sciences and Research*, *4*(5), 1692–1708.

Feng, X. Z., & Liu, J. Q. (2014). Monitoring on mode selection for technology transfer based on regional sustainable development. *Applied Mechanics and Materials, 513-517*, 3864–3867. doi:10.4028/www.scientific.net/AMM.513-517.3864

Feng, X. Z., & Nie, Q. (2014). A monitoring model of rebound effecting for technology transfer. *Applied Mechanics and Materials, 543-547*, 904–907. doi:10.4028/www.scientific.net/AMM.543-547.904

Jahn, R., Muller, O., & Bozorgmehr, K. (2015). Characteristics and determinants of knowledge transfer policies at universities and public institutions in medical research – protocol for a systematic review of the qualitative research literature. *Systematic Review, 4*(110), 1–5. PMID:26286398

Jian, L., & Yu, H. (2014). The restriction mechanisms on R&D result transfer performance for Chinese universities based on Grey incidence analysis and Optimized GM(1,1). *Journal of Grey System, 26*(3), 12–22.

Koschatzky, K. (2002). Networking and knowledge transfer between research and industry in transition countries: Empirical evidence from the Slovenian innovation system. *The Journal of Technology Transfer, 27*(1), 27–38. doi:10.1023/A:1013192402977

Lavhengwa, T. J., van der Walt, J. S., & Lavhengwa, E. M. (2014). Factors influencing e-collaboration for knowledge development and innovation. *SA Journal of Information Management, 16*(1), 1–8. doi:10.4102/sajim.v16i1.588

Lundquist, G. (2003). A rich vision of technology transfer technology value management. *The Journal of Technology Transfer, 28*(3/4), 265–284. doi:10.1023/A:1024949029313

Nelsen, L. L. (2004). A US perspective on technology transfer: The changing role of the university. *Nature Reviews. Molecular Cell Biology, 5*(3), 1–5. doi:10.1038/nrm1337 PMID:14991004

Nguyen, N. T. D., & Aoyama, A. (2015). The impact of cultural differences on technology transfer. *Journal of Manufacturing Technology Management, 26*(7), 926–954. doi:10.1108/JMTM-09-2013-0130

Shah, G., Kamble, P., Shinde, R., & Ghatge, K. (2014). Advances in technology transfer for the Pharmaceutical industries: An overview. *Research Journal of Pharma and Technology, 7*(6), 719–726.

Smilor, R. W., & Gibson, D. V. (1991). Accelerating technology transfer in R&D consortia. *Research Technology Management*, *34*, 44–49.

Tu, J. (2012). Die and mould technology transfer in the environment of industrial cluster. *Applied Mechanics and Materials*, *201-202*, 1029–1032. doi:10.4028/www.scientific.net/AMM.201-202.1029

Woerter, M. (2012). Technology proximity between firms and universities and technology transfer. *The Journal of Technology Transfer*, *37*(6), 828–866. doi:10.1007/s10961-011-9207-x

ENDNOTES

1 Good Manufacturing Process – part of the requirements in drug approval.

2 Referring to the handgun with a revolving bullet case.

Chapter 5
R&D Support Program Types

ABSTRACT

In this chapter, the R&D support program is discussed, it's characteristics and needs as well as the different roles the different players play. The description is general, followed by some of the different characteristics and types of programs. To better illustrate some of the points discussed a specific sector is analyzed as an example. The structure as well as the basic guidelines for establishing and R&D support program are discussed. The R&D program is defined as is translational research followed by guidelines for the operation of a program and ending with additional recommendations.

BIOMEDICAL RESEARCH AN EXAMPLE OF A SECTOR

For an illustration of the potential of an R&D support program we need to establish first the understanding of an R&D process. In this section, we shall discuss the R&D process of a drug. There are very few programs that deal with the entire process and therefore it is important to identify which programs support which steps. The description is simplistic and serves as an illustration of the process.

Phase 1: Basic Research

The basic research can start from different starting points:

DOI: 10.4018/978-1-5225-3652-9.ch005

1. Researching a specific disease and trying to identify the chemical biological cycle responsible for it. A potential result is an understanding of the chemical cycle operating the disease with the different compounds and cellular organs involved in it. These offer potential targets for future drugs to either stop the process, repair it or prevent it;
2. Testing different compounds for stopping a certain process that is known regarding a disease, original compounds either natural or synthetic – originating from libraries or from traditional medicine, these are potential hits;
3. Accidental events happening in the lab – recall how penicillin was discovered?

Once either target or hit are suspected, the research will focus on two major aspects, to try and show that the target is valid, or that the hit has at least some preliminary effect (at least a specific minimum doze shows some impact); and a better understanding of the process in order to be able to further the work. Most of the basic research support programs / grants finish at this point their support. Safety is dealt with at any of the preliminary steps, as if the drug candidate is potentially harmful the rest of the work is normally useless.

Phase 2: Translational Research

Following the above some medical chemistry follows in which hits for the known targets are designed. Once that is done or if the hit was the starting point, the next step will be a better design of the hits, defined as improved efficacy by reducing the doze required for efficient proof of concept. In many cases this work can be done using cell and tissue cultures. It will also require some sort of anti-toxicity proof, unless the compounds are already known as nontoxic. During this stage, a small number of candidate molecules are selected (normally one or two) as the best candidates. This is done normally using several rounds of medicinal chemistry, developing "families" of molecules based on the most promising hits and testing them. The aim of this process as mentioned above is to arrive at the end with the most promising molecule to continue the process with. Once that is done the move to the next step is performed.

Phase 3: Pre-Clinical Research

The next step in this model process would be the preclinical trials, preferably performed on an animal model developed to simulate the researched disease. The development of such models can be complicated, and there is of course a preference for the proved and accepted models which require no development, but still require setting up and validating. In some cases, no accepted model exists, or most models are not considered good enough. One should remember that the model should simulate the disease in the animals to allow for the cure to be developed. One should also recall that this phase is a step towards clinical trials in humans. Pre-clinical trials will include anti-toxicity if it had not performed before, as well as several animal models (increasing in size if required), depending on the disease.

Phase 4: Clinical Trials

Clinical trials on their different sub-stages are included in the domain where firms intervene as well as product support programs, where they can offer enough support. Under these stages the development of the drug including the production process and regulatory approval are performed.

The process as far as we are interested, stops here with the approval, hopefully of the drug candidate as a drug including the production process. One should remember that this stage is within the specific domain of the industry.

The process described above is a general process the starting point of which may vary. Some of the steps may change and may be performed more than once. A molecule, "hit" may be used to validate a target, but another molecule may later be selected or developed as a better "hit" for further development into the drug finally approved. Known drugs may be approved for additional diseases, or many other variations. The process described above is intended to allow us to discuss the support programs. The process in most cases entails a start (basic research) by the academy with the research performance transferring to the industry towards the clinical trials.

The R&D support programs can support the process in most of the phases described above, even though the costs of the last stages of the clinical trials, especially stage 3 can be very expensive and more than an R&D support program could or would like to support.

Here again one should recognize that as the academy rarely works according to the standards required by the regulatory authority, any stages submitted to the regulatory authority as part of the approval process would need to be performed under the relevant standard demands, and therefore some phases may need to be performed more than once.

The R&D support program may support the academy in its early phases, the joint work of the academy with the industry or the industrial phases. The role of the program will be described below, but it is important to recall that they are normally defined to aid in one phase or two, and therefore deal with specific entities, academy, industry or combination, and will define the eligible entity or entities for support to fit the phase it intends to support.

SPECIFIC ROLE OF TRANSLATIONAL RESEARCH

The translational research can be presented as the basis for technology transfer and also as the basic move towards the market from the basic research (Bell et al., 2011; Zhang et al., 2011; Talman, 2012; Pati et al., 2013). Translational Research as described on page 1 by Duet et al. (2010) ". The progression of scientific advance from inception to clinical practice…" was evidently first defined in the field of medical research to denote the move from basic research to the clinical phases. It has however progressed into other fields and it evident in all technological fields (Carr, and Bradley- Levine, 2016).

The role of translational research in the research and development process was described above. The discussion at this point regarding translational research shall focus on the role it plays regarding technology transfer.

Since translational research translates the results of basic research into a potential application, the nature of the process, is one of transit. The transit is also a mark of a move from an activity that is entirely academic to one that is potentially in the combined zone of cooperation between academy and industry. The move of the focal point towards application and thus the balance shift towards the industry, will require eventually a move of the technology itself, ownership and activity of further development to the industry. Thus, the phase of translational research requires if not an actual transfer of the knowledge, at least preparation towards such a move, which shall be required in future phases.

Basic research, curiosity driven for the better understanding of the world and the laws of nature, is so far removed from application in most cases that the potential applications in most cases are difficult to foresee and even the relevant industry is not clearly identified. Therefore, the preparation towards transfer of the knowledge are not relevant. However, in the translational research phase, the application is defined, and thus automatically the potential sector, industry and in some cases a group of firms. Once the sector and industry are identified, the future demands regarding the IP can be taken into account, and thus the project can be designed to prepare for that. Following that, the program when dealing with such projects and demands, can be designed to cope with the IP demands in the by-laws.

WHAT IS AN R&D SUPPORT PROGRAM?

An R&D support program is a financial scheme that finances at least partly R&D projects, and that may offer additional means of R&D support (e.g. mentoring, partnering assistance etc.).

As the definition above mentions, these programs are mainly financial tools. Whether governmental or private, they aim to finance the research and development as they define it, and thus to aid the execution of research and development projects. Therefore, an important factor when analyzing such programs is the financial aspect of the support given. That is to say, the amount of support, the rate of the support and the type of the support – loan or grant.

It is important to note that all additional means of support offered are facilitators for the programs and meant to aid the specific type of the projects supported by offering answers to additional needs, but that these are not the common support in all the programs and are in some cases neglected.

The financial schemes are considered strategic, as is discernible in their normally long-term view and strategic intentions, mainly focused on long term economic development.

To sum, the R&D support program is a financial scheme or tool with the strategic goal of economic development among others, offering in some cases additional support measures to its supported R&D projects.

DESIRED IMPACT

The desired impact for each program is the reason-d'etre for the program, the reason for which it was established and funded, and which if it can show that it is progressing in the direction of that goal would probably continue to be funded.

It is important to be able to define the desired impact in quantifiable terms. Specific rather than general terms are preferred.

A program that would seek to increase significantly the number of small to medium sized enterprises (SMEs) in a region, would be harder to assess than a program that seeks to double the number of SMEs in the region. The second requires that one measures the number of firms at the start and end and see if the growth complies with the goal. The first would require a definition of significant, which would tend to be a clear-cut term, having different meanings for different persons.

When defining the impact, it is of course important to make sure that the measurement at the end would comply with the desired result, even if that requires some sort of measurement effort at the start as well. So, when designing an R&D support program and defining its' goals, one should bear that in mind and install the measurability into the goals.

It is important to recall that the desired impact may be complex as well as diverse, it can include economic and social aspects as well as aspects related to education and qualification of the work force. Not all of the goals or the desired impact aspects need to be as important, there can be primary goals as well as secondary goals, and the desired impact should consider the secondary impact just as well as the primary. Especially in cases when the secondary impact may have undesirable results. During the high-tech bubble days in the last 1990's and early 2000 firms participating in R&D consortia discovered that the cooperation in the R&D projects was a good opportunity for evaluation followed by recruitment of evaluated personnel from other firms in the consortium or from the academy. That was a secondary impact and one which could have reduced the tendency of small or less attractive firms from participating in collaborative projects, for fear of losing their personnel. When such unwanted phenomena raise its head, it is up to the program to find a solution across the board and make sure that the program as a whole is not negatively affected. A potential solution in such a case would be to have the firm commit to an ethical form of behavior combined with an enforcement mechanism (e.g. ASTP-Proton).

Before moving to the different types of goals themselves, one more point to recall regarding the desired impact, and that is the time element.

The desired impact may take some time to become evident. If you want to create a growth in the agriculture sector it would require probably a period of R&D and several periods or agriculture cycles for the desired research results not only to become known and to be adopted, but rather for the end results implemented to make the required effect. If dealing with

Pharmaceutical drugs the time required for the drug candidates or become approved drugs in the market and to show relevant revenue stream may be over 15 years, and for a whole group it can be around 20 years. That is to say that measuring the efficacy of the R&D support program may not be as easy or as quick as may seem at first, and it is recommended to term the program if necessary accordingly. It may be more advisable for the program to be defined interminably with periodical assessments to check its progress towards the end goal[1].

Economic Growth

Economic growth, and more specifically the contribution of different factors to it, is problematic to measure when dealing with on-going activities that can benefit not only from different measures but also from the combined effect of several contributing factors. If a firm presents a significant growth in the sales, that wonderful result may stem from the improvement in the sales, from the increased efficiency of the production, the improved financial management that allowed for better deals on its products, or perhaps from all of them together, making the identification of the specific contribution of each factor very difficult.

Improvement directly due to sales is relatively easy to monitor as it is close enough to the end of the value chain to be directly measured in units or in $. However, the improvements in the products, the results of the R&D effort are more difficult to measure in a "pure" way as they are far enough up the value chain to be combined with other potential contributions to the improvement. Additionally, it is difficult to measure the exact contribution compared to other parameters, the reputation of the older products, improved sales personnel familiarity with the products etc. and therefore it is much harder to justify the effort in this way.

The above refers to efforts internal to the firm. When dealing with external support, as are all R&D support programs dealt with here, it is much more difficult. This parameter has some influence over the preference of programs in their support, and on the definition of success. There is also a tendency to see the completion of the project, as per the goals set to it at the beginning as a success of the project. That is due mostly to the nature of some of the programs we shall discuss below which do not reach the market with a product and therefore seek a more "local" well defined end for the activities.

Most programs would seek secondary goals that are contributing to economic growth, but that can be measured in a more specific way and therefore serve as a more direct index for the success or impact of the program. That substitution of more easily measurable results instead of the actual more difficult to measure results while natural in the end may tend to obscure the real goals of the programs. Programs may tend to see in time their role as supporting or financing the industry and forget that it has a higher goal in economic development.

Non-Economic Growth

Looking at the above, one would be justified to assume that all goals are economically aimed. However, that is not entirely true.

R&D support programs may be aimed at the development of strategic capabilities, aimed at the strategic safety and security of the national entity. Development need not be limited to weapons directly, it may also be related to issue that will allow the local industry to function without dependence on external support. Or can be aimed at independence from constraints, economic, political and others. The effort directed at the development of alternative fuels for transportation (alternative to fossil fuels) can be seen as such a move. While not directly aimed at economic development, and certainly in a country like the US which has its own fuel resources, not necessary for economic development, such a move to free the US economy from dependence on external fuels is strategic.

There are ample examples for military oriented R&D that is first and for most concerned with the national defense and only in second place in economics – export. That is evident when checking the list of technologies, the US forbids export to the countries it considers a threat, and easily recognizable when discussing nuclear capabilities. The US has gone to war to stop the development of such capabilities in Iraq.

So, while some R&D programs may have a secondary economic goal, their main goal is strategic and economics come second.

GENERAL OPERATION PROCESS OF AN R&D SUPPORT PROGRAM

While the programs may differ in the operation process which is defined to fit the strategic goal of the programs, since they are financial schemes of support for R&D projects, there are some basics in their rules, that all must follow. The general process is described below:

Step Zero: Preparations

While it may seem that the preparation for the operation of the program is not part of the process, that impression is a mistake. The R&D support program is an action taken by a governmental entity in order to support a commercial or research entity active in the market. That means that the entity forming the regulations and process is doing it for an entity operating in a completely foreign environment and with difficulties and needs that the governmental entity can only guess. Therefore, the preparations should be seen rather as an on-going process of learning and improving the R&D Support program, adapting it to emergent needs and thus as an on-going activity in parallel to the operation. This is the reason for naming it a step, but still referring to it as step zero.

For our discussion here, it shall be divided into the basis required in order to commence operations, and the on-going learning process (step by step).

Basis for Operation

The basis for operation of an R&D Support program should include the following:

- Regulations for application to the program – including the time tables, forms and required information checklist, accompanied with whatever clarifications are required.
- Regulations and by-laws for the operation and closing of the projects under the program – making it clear how the projects will be contracted, managed, monitored and finalized, and any instructions to survive the end of the project. These will refer to administration, budgeting and IP at least, but may include additional data.

- Clear financial instructions and regulations as an item by itself, which can be sent to the relevant units in the potential applicants and grant receivers.
- Description of the application, evaluation and granting process, including a copy of the future contracts where applicable.
- Address and relevant personal for future questions and requests.

While the list above may seem daunting it is necessary in order to assure that there are applicants ready to apply.

Eligibility criterion should be clear and easy to understand. The application of only eligible parties is in the interest of both the applicants (no one wants to waste time), as well as the government entity managing the program, that does not want to waste time and effort in dealing with proposals that are not eligible. The clear, and just as important clear-cut eligibility criteria are very important can as a secondary measure help in raising the contentment level of the applicants from the program.

Eligibility check list:

- Clearly setting the parameters both for proposals and applicants;
- Setting rules for unforeseen cases – basic rules that can be applied to test unforeseen cases, as a measure for future development;
- Short – lengthy description give rise to misunderstandings;
- Setting the model proposal and applicant as an example;

Clear financial guidelines for setting the budget, including eligible costs and a clear list or definition of ineligible costs. This is very important in order to make sure that the applicants know what to ask for and know what they are getting into financially. While it may seem the research organizations, being non-for-profit organizations in many cases, are not interested much in the budget, this is a mistake. Research organizations as well as non-for profit other organizations, are commercial entities just as are firms. The fact that they do not seek profit does not mean that they are not controlled by budgets, and that therefore they are just as financially aware as are firms. They need to know what they are getting into, as in most cases, unlike firms, they do not have other financial resources available for covering mistakes. The importance of the financial guidelines should be extended also to the management of the project, as well as to the closure of the project and final settling of the accounts. Once again, an organization that mismanages the project through lack of understanding may rue the day it won the grant, as the damage is greater than it would have been, had it not received the grant.

Example 1: Financial Guidelines for Equipment

A university won a grant from a program that allows for the purchase of equipment, but finances it by recognizing the depreciation on the equipment. The equipment is depreciated over three years. The university, happy with the new grant started the project, and after six months ordered the equipment which was delivered and installed (invoice issued) six months later. When coming to settle the accounts the grant approved only 67% of the cost – as the equipment served the project for only two years. The university had to finance that cost, which it did not plan to. The guideline stated that the depreciation is calculated from the installation or invoice date, whichever is the latest and only up to the end of the project period. Had the university been aware of that, it would have ordered the equipment much earlier, and received full reimbursement.

One of the major issues regarding the operation of the grant is the ability to make changes to the work plan and therefore to the budget. It must be noted, that if the budget is derived from the work plan, as the finance required for its execution, any changes to the work plan should have some impact on the budget – and vice versa. The process for requesting such changes as well as the time for making them, the approval process and response time should be made clear from the offset. The bigger the technological uncertainty of the projects under the program, the greater the probability that changes will be required.

Of course, if the financing is done by "block payment, as may happen in academic funds, meaning that the grant recipient receives the funds entirely at the start and has to report the results and expanses, the case is different. This, however, is not the case with most industry supported R&D support programs.

IP issues are very important to all participants and should therefore be made clear from the beginning. There are several reasons for making the IP rules clear, and for doing it early:

- Firms and lately and to a growing extent research organizations are sensitive to the ownership and usage rights of knowledge, and therefore want to know what they can, or cannot do with it at the end of the project, and if any costs will be incurred per action;

- The main purpose of the R&D Support program is the creation, transfer or adoption of IP or knowledge and therefore the rules for doing that are important for the government entity managing the program. In many cases both what it wants to have happen or what it absolutely does not want to happen are very clear to it, and should be clear to the potential applicants before they make their move;
- For high-tech and startups, the IP is of critical importance, and in cases of uncertainty as to what will happen, or if they have to share their existing IP as well as future one, they may not wish to get involved in the program at all. The value of such firms may depend on their ownership of IP and if that is compromised, or their exclusivity reduced, these firms may show a decrease in value – which fear alone is a deterrent to taking up such programs.

Step One: Call for Proposals

In order to support R&D project each program must generate a pipeline of proposals which it will process later on and support at least part of the proposals, which will thus be turned into projects. In order to generate that proposals each program must issue a call for proposals stating what kind of proposals / projects it expects to support. That call for proposals can take various forms: it can be a sub-case of a call for tenders, issuing a detailed description of the R&D to be performed, with expected proposals to compete mostly regarding the cost of the R&D project; it can be a more open call with a general statement regarding the expected project or impact, allowing the proposer to present a proposal with its own scope of work, specific expected results and cost. These, more open-ended proposals are normally less detailed when issued.

The calls also differ in another major aspect. Some of the calls are issued specifically at specific times and are normally issued only once for a specific scope of work. Others are issued on a general basis, mainly, they are open calls, which means that proposals are collected up to a certain fixed date (normally annually) evaluated, and some of the proposals are approved for funding. Proposals arriving after the pre-fixed date, wait for the next evaluation date. The governing entity has a vested interest in generating proposals, as these are the basis for the project which are the initial aim of the entire program, and without enough proposals and project the program would in most cases be cancelled.

Step Two: Evaluation and Financial Support Award

Once proposals are submitted they have to be evaluated and a decision regarding which proposal is to be financed must be made. Here the criteria for evaluation, normally published with the call for proposals needs to be applied in the evaluation of each proposal. A lot has been written about the evaluation of proposals and the award process. However, it is normally comprised of the following steps:

1. Eligibility check – checking that each of the submitted proposals is eligible for funding under the rules and regulations of the support program as published in the call for proposals or the by-laws of the program. The funding agency is normally conservative regarding this point, making sure that only the approved targeted sectors are getting the support;
2. Technical evaluation of the proposals in which the goals of the proposed project as well as the technical back ground, and the proposer's ability to perform the proposed research are evaluated. While the previous step can be performed by administrators as the eligibility check is normally administrative, this step, the technical evaluation needs to be performed by technical experts, which the program has to find and fund;
3. Once the technical evaluation has finished and all the different proposals have been graded, a decision regarding the funding of each proposal needs to be taken. This step can be performed by the administrators, the technical experts or some combination of both. In most cases a committee is established[2];
4. Announcement and contracting – once a decision has been taken an announcement and contracting follows. The decisions regarding the proposals chosen for funding need to be forwarded to the proposers, and as the award includes an agreement to fund a specific project this action has to be done formally, especially when dealing with public funds.

Step Three: Project Management Post Award

Once the projects have been approved and financing granted according to the program's regulations, the projects move to the execution phase. This phase is performed by the proposers, and at best the program keeps track of both the finance involved and the work progress. Some programs are more involved in the projects, demanding monthly or quarterly reports, while others are less involved, demanding a yearly or project summary reports only. This is by far the most common activity in an R&D support program, as it last through the entire duration of the program and sometimes extends long after the program has finished awarding the funds, and issue no more call – see the cases of the EU framework programs for research which continued for years after the framework has finished allocating the funds, as the projects were approved for three or five years.

It is therefore important to note how the program manages the projects it approves and what will be the future demands from these projects, that can extend beyond the duration of the program and even the projects (e.g. s limitations on the transfer of the knowledge). Once again, the monitoring of such projects is normally done on the two most extensive levels, the expenditure rate and technical performance, according to the approved budget, and the advance in the technology. While expenditure reports can be generated automatically if the awarded entity performing the R&D project is well managed, the progress reports are more difficult to assess in some cases, and require once more the involvement of experts in that assessment. The reporting is regarded in many cases by the awarded entity and the administration of the program both, as necessary evil. However, the technical progress report is a very important strategic management tool. In many cases the knowledge that such a report will have to be submitted in the future will help the R&D team focus their work on the proposed and approved R&D work-plan, and prevent deviations. As the work-plan is approved, at least in part, according to its compliance with the strategic objectives of the R&D support program, adhering to it, means adhering to the strategic goals of the R&D support program. It is important to stress the point that the technical report is not just a monitoring tool by allowing the program administration to check on the progress, but also before the act as a tool to keep the R&D team focused on the goals of the project and the program.

Detailed Aspects

National and international R&D support programs need to predefine most of the rules and make sure that as much as possible is clear, the reason behind the need for clarity, and other critical points.

The aspect of nationality, or rather the regional definition of the programs is crucial. As mentioned above in the desired impact sub section, one of the goals of all R&D programs is the eventual economic development, in a specific region. The definition of the region relates to two aspects: the source of the funding and the eligibility of the applicants. When we are dealing with a program that is limited regionally, that is normally derived from the source of the funding. The easiest to understand is a national program that is operated based on a national budget source being fed from the national public funding. This however, is the simplest case, and as such is demonstrated in many general programs.

However, more specific programs, dealing with specific needs of different reasons, and their existing resources, building on the capabilities existing already there, would normally be the result of a special need. Focused on specific problems in creating economic growth in that specific region. Such programs would tend to run into difficulties when they are being funded from the general budget of the nation, as other regions would be sponsoring the specific region at the expanse of their own development. It is possible to reach a compromise that would allow entities from all over the country to apply, but the projects themselves would have to be carried out within the required region.

There is a possibility to reach another compromise and that is to operate a program all over the country, but vary the details, or at least the financial support details per region.

Examples can be found in many countries, either supporting more rural areas to become more industrialized, areas with more traditional industries to become more high-tech (e.g. the effort being done in Rotterdam to renovate the industrial profile of the harbor area). The tests for determining the residency of applicants should be clearly stated in the publications of the program.

Methodology

The finer rules and regulations should be developed in coherence with the purpose of the program. While it is natural to expect a program that intends to encourage the migration of academic R&D human resources from Academy to the industry to support mainly salaries, and perhaps some of the additional costs of the research to be performed during the integration of the new workforce in the industry, equipment is another story. One would expect that the industry would have most of the equipment required for such work, and the program would be justified to support the purchase of new equipment only marginally.

While other expanses supporting participation in workshops and exhibitions and even publications would seem more coherent regarding the final aims of the program. However, finance is not the first step, the first step would be to define the program goals (preferred quantifiable).

Goals

As we discussed above the main issue would be to set the goals for the program. There are different goals that can be ascribed to the programs: political, social economic, or even external pressure. Only once the goals are set can the other steps follow. The finance (discussed below) would need to be developed in a coherent way so it is compatible with the goals. It is for this reason that copying programs from one country to another can be very risky. The goal setting for a specific program is country or region dependent. It needs to take into account the entities active in the region, and the infrastructure that the region can sustain. As these differ from region to region or among countries so the goals would need to be modified.

Example 2: Two Programs With the Same Aim

Two programs that aimed at introducing academically educated human resources into the industry in two different environments were therefore designed differently. The Danish Industrial PhD program (see here: https://innovationsfonden.dk/en/application/erhvervsphd) which wanted to introduce PhD holders into R&D positions in the industry, working in an environment where there was room for more PhD holders, opted to allow people while working in R&D projects for the industry to receive their PhD in partnership with an academic institute. Accepting that the graduates of the program do not intend to develop an academic career. The Spanish Torres Quevedo[3] program which wanted also to introduce PhD holders into R&D positions in the industry, operating in an environment where there were many PhD holders that needed to be "relocated" from the academy to the industry, supported their employment post PhD receipt, in the Industry by financing a decreasing part of their salary costs as per their employment in the industry. The existence or lack of it of PhD holders determined both the goals and therefore the type of finance offered under each of the programs.

Finance

The financial regulations refer often to two aspects, the rate or type of support given, and the types of expanses that are reimbursed or supported (this topic was discussed above - the following is a summary).

Types of support can be categorized as follows:

1. **Funding:** Funds are transferred to the parties executing the projects under the program to enable the execution of said projects;
    a. **Grants/Prizes:** A lump sum is supplied to the parties (jointly or separately) executing the projects and they can exercise freedom in utilizing the funds without reporting the expenses incurred (though some limitations or taxation may be incurred);
    b. **Loans:** The funds are delivered on condition that they be reimbursed to the funding agency when the project results bear revenues, or at another point in the future;
    c. **Reimbursement:** Wholly or partially reimbursing the parties executing the projects for the expenses incurred during the project life;
    d. Combination of two or more of the above.
2. **Guarantees:** No direct funds are transferred to the parties executing the projects but guarantees in various forms are supplied in order to allow the executors of the projects to raise their own funds under terms more favorable than the normal market terms
3. Combination of funding and guarantees.

The main differences regarding the two major types (not the combination) is that while the funding requires allocating a budget that will be used for funding the projects, either per year or per specific duration, and which can be liable for cuts or increases therefor introducing an element of uncertainty regarding the program (e.g. the EU framework programs that each time when a new budget is form change their operation and some of their regulations and funding mechanisms). The guarantees while not requiring a budget allocation as long as the normal budget can cover any incidents, may not be liable to the same uncertainty mentioned for the funding schemes, but are also normally capped, and more important, require a strong enough financial sector or private resources for funding to supply the actual funds required for the execution of the projects. The use of combination normally is not intended to negate the disadvantages of the above, as normally it results in negating the advantages and not the disadvantages, but rather to allow more potential clients, with different levels of government intervention phobia to participate.

The "classical" categories of expanses to be reimbursed can be classified as followed:

- **Human Resources:** Salaries (including social costs), scholarships etc. as directly linked to the projects supported.
- **Sub-Contractors:** Including consultants and ranging from technical to social economic and others, either permanent or per task in the project.
- **Materials:** All consumables that are consumed (cannot be used after the project) during the project duration.
- **Equipment:** All other materials, components, instruments and equipment that are not consumed (can be used after the project ends).
- **Legal and Patents:** These are normally separated from the other types of consultants as these are both often critical for the success of the technology transfer and utilization and as they often include both the sub-contractor's fees as well as taxes and submission costs.

There are many more categories and sub-categories that may arise – travel costs for projects that involve international cooperation, relocation costs in projects that involve the mobility of personnel across countries and publication as well as other costs. These are normally grouped under "miscellaneous" or "others", unless they are directly aimed at promoting one of the project goals.

There is one category that is normally problematic when dealing with reimbursement, and that is the overhead. There are many accounting systems and explanations regarding the topic of overhead, or indirect costs and unless the legal system requires that the program adopt a specific single methodology for computing the overhead, many programs would rather avoid it by either indexing it to another category of expanses (Human resources, material etc.) or by determining a lump sum relevant to the project size. The actual calculation is cumbersome both for the funding agency which is required to check the reporting it gets, and for the parties executing the projects required to monitor and calculate in a substantiated way the overhead. It is possible to lay the burden of proof at the applicant's door, however if it complicated it may not be worth the effort to invest in the infrastructure for a single project, reducing the take-up.

Legal

While this may have been deemed to be the most important item in the methodology for developing the finer points of the program, this is in fact a simpler matter than the finance. The legal environment can in many cases be solved by legislation, which in many cases would be required at some level due to the finance aspect. Once the finance aspect is determined (type of parties to execute the projects, strength and diversity of financial sector, involvement of foreign entities etc.), and the program goals are set, there remains just to identify name or establish the managing agency and prepare the legislation to allow it to execute and monitor the program.

However, there can be important legal aspects that would influence the take-up of a program. Under the German legal system, an exclusive license granted by the academy to a firm to utilize an academic invention would be considered as a valuable possession of the company, and upon dissolution of the firm would be counted as a property of the firm, and be disposed of as the other property items are. That would mean that a university in Germany would need to consider that by licensing its technology to a start-up firm, it is partaking in the risk inherent in such a young and vulnerable firm. Under the English legal system, the university could grant an exclusive license to the firm but stipulate that in case of dissolution the license will be revoked. That specific caveat would allow universities easier licensing as they are not part of the risk takers on behalf of the firm, other than risking potential revenues. While the German system may allow the firms and universities to find specific loopholes, these would be the special cases, and therefore would eventually reduce the tendency of the universities to license to start-ups, compared to other systems.

Another example from the aspect of tax legislation could the Israeli Angels law. That law allows investors under certain conditions (both on them and on the invested firm) to invest in start-ups focused on R&D, and have that investment considered as an expanse / loss, for the financial year in which they made the investment. The rational is, that as most such firms fail, it is more encouraging for the investor to recognize the loss at the time of the investment, and if there is success tax the whole revenue at the point in time when this will happen, but form the moment realize the loss and offset it against other gains, than to wait for the firm to either succeed and then tax the difference. Or if it fails to recognize the loss only in the future. This has several advantages for the tax authorities as well. They no longer have to see proof that the invested firm has dissolved, or if it is not dissolved, cannot repay the investment, but rather just wait for the investor to realize his investment and then have it taxed to the full. Otherwise for them it has already been off-set against revenues in the past[4]. For the investors, that recognition makes proving the loss redundant, and as this loss can be off-set against all other revenues, serves as a tax benefit.

For the high-tech firms, this is a potential advantage for the competition with other investment options, and allows more private funds to be diverted towards investment in the high-tech sector.

Investors in the early stages of development of a new venture in high-tech, called "Angels,“ as they save the new venture when it is still very young and fragile, and therefore are somewhere between regular investors and philanthropists. However, this is not pure philanthropy, therefore it still needs to have some business outlook, and to encourage that activity the tax deduction serves well.

So far, we have seen two aspects of legislation that could have an impact on the program both regarding the design and the execution or take up of the program by the eligible applicants. Other important legal aspects would include the IP laws and regulations, the employment and social benefit rules in general as well as more sector specific regulatory issues. Since we discussed the IP legal aspects in other sections of this book, at this point let us discuss the employment aspects and their potential impact on the R&D support programs.

Since R&D project management, as a rule entail employing personnel to perform the R&D projects themselves. The legal environment regarding the employment of personnel is relevant. In most countries, there are three categories of employment, direct employees that are hired by the applicant and that work under a contract according to the relevant laws of the land. The second major category is sub-contractors, which may be consultant or other workers employed for a specific task only and are normally free to engage with other entities regarding work. They would have their own business and maybe employed for the project benefit either per task (globally) or per hour/day/week but are not considered employees of the applicant and are legally independent.

The sub-contractors, as non-employees may work outside the premises of the applicant and the applicant will exercise only nominal and general control over them. They are required to perform tasks and deliver deliverables, but manage their own process to a large extent. Therefore, when designing a program and demanding from the applicant/executor of the project to substantiate the expanses, for the employee expanses it is possible to demand substantiation from the applicant / executor, but from the contractor only the output, and actual payment can be substantiated by the applicant / executor. Therefore, when allowing the employment of sub-contractor, one should take into account that the executor of the project would have little control over their actual work and in most cases, be able to measure only the outcomes and expanses relevant to their employment. One should also bear in mind that under some legal systems sub-contractors may become employees under certain terms. Therefore, the program should make sure that it is not liable for any demands or law suites as a result from such a shift in the status of a sub-contractor in one of the projects supported.

Most programs feel uncomfortable regarding sub-contractors from another reason. The end result of an R&D project is knowledge. In the approval process, it is clearly the intention of the program that the knowledge resulting from the project will reside with the applicant approved to execute the project, and not some third party not under direct obligation to the program, and one which in many cases has not been named, and of course thus not evaluated when granting the funding. Therefore, it is normal to see some limitation in the scope of sub-contracting allowed in projects under R&D support programs, either by extent of the work, of the budget or type of work allowed (core technology). There may also be demands on specific terms to be included in the contracting of the sub-contractors (e.g. ownership of the knowledge resulting).

There is a third category, which is in fact a sub-category of the employees and that exists in the academic world alone – scholarship receivers. A scholarship grant is in fact a sort of salary, but not including many of the regulate social benefits included in salaries and therefore also in most cases exempt from taxation.

Scholarships are intended to be short termed covering a specific phase in the academic life, that of a research student. As such they are intended for young people and therefore are bereft of many social aspects (e.g. pension saving, health insurance, supplementary etc.) and represent a sort of reduced short-term employment salary. This is a sub-category of employment as they work on the premises of the applicant / executor, and they direction and instructions of his workers, as well as being directly paid by him, and being obliged to him in additional ways. Regarding the demands as per reporting on their work, while this is clearly within the power of the applicant / executor, it would seem to be demanding too much as they normally have other obligations under the applicant (teaching, studying etc.) and their cost to the applicant and to the program is minimal.

However, just like in the case of sub-contractor it is important for the program to be able to protect itself from demands arising from a status change into employment which may arise due or during a project under the program (the EU framework demanded that scholarships under the Marie Curie programs contain all social supplements, and that the applicants commit to undertake any demands in the future arising from such projects regarding salaries or social benefits).

Other forms of employment such as via employment agencies which are a sort of sub-contracting, affiliation from other units etc. are considered as sub cases of the above.

It is important in that legal aspect of employment to remember another aspect and that is the aspect of the ownership of inventions or ideas resulting from the work. Legal systems view such cases differently, from systems that deem all knowledge as owned by the employer, to others that deem the knowledge to be owned by the inventor alone. It is important to address these issues when drafting the employment contracts as well as when contracting sub-contractors. The program as a general measure can demand in its by-laws that such actions be taken, either adding text predefined by the program or by demanding submission of the relevant contractors for approval.

Specific sectorial regulations are regulations that are required by specific sectors either for the introduction of projects, products or employment. The most known are the regulation regarding the approval of pharmaceutical drugs operated by the US FDA. While it may be clear that programs that support the development of pharmaceutical drugs up to the final lunch into the market should consider the regulatory demands as well as the cost of the FDA approval process, it should also be clear that even for programs dealing only with the earlier stages of development of such drugs, the FDA regulatory would have an impact.

Good Laboratory Practice, referring to the drug development stages at the lab preceding Good Manufacturing Practice is therefore an early demand and may impose requirement on the R&D process which the support program would be wise to take into account. Other examples relate to the communication standards in developing components, and numerous regulatory issues that are relevant for programs aiming to look at the whole R&D process, or to prepare for it.

REFERENCES

Bell, E., Harpur, S., Doherty, K., Struber, J., & Davis, L. (2011). What is Translational Research? Background, concepts and a definition. *International Public Health Journal.*, *3*(2), 133–143.

Carr, K. M., & Bradley-Levine, J. S. (2016). Translational Research design: Collaborating with stakeholders for program evaluation. *Qualitative Report*, *21*(1), 44–58.

Duet, L. J., Preedy, D., Thomas, V., & Cree, I. A. (2010). An exploratory investigation of the influence of publication on translational medicine research. *Journal of Translational Medicine*, *8*(62), 1–5. PMID:20064266

Pati, D., Pietrzak, M. P., Harvey, T.E., Jr., Armstrong, W. B., Clarke, R., Weissman, N. J., ... & Collins, J. M. (2013). Advancing translational Research through facility design in non-AMC hospitals. *Health Environments research & Design journal, 6(3)*, 126-137.

Talman, W. T. (2012). Views of Translational Research from a somewhat Translational Scientist. *Journal of Investigative Medicine*, *60*(6), 863–868. doi:10.2310/JIM.0b013e31825fd8b8 PMID:22781556

Zhang, Z. Q., Chen, H. Z., & Liu, D. P. (2011). Translational research: Lessons from past research, growing up npwadays, and development goal in the future. *Science China Life Sciences*, *54*(12), 1085–1088. doi:10.1007/s11427-011-4252-0 PMID:22227899

ADDITIONAL READING

Hobin, J. A., Deschamps, A. M., Bockman, R., Cohen, S., Dechow, P., Eng, C., & Galbraith, R. et al. (2012). Engaging basic scientists in translational research: Identifying opportunities, overcoming obstacles. *Journal of Translational Medicine*, *10*(72), 1–14. PMID:22500917

Main, D.S., Felzien, M.C., Magid, D.J., Calonge, B.N., O'Brien, R.A., Kempe, A., and Nearing, k. (2012). A community translational research pilot grants program to facilitate community-Academic partnerships: lessons from Colorado's clinical Translational science awards. *Progress in community health partnerships: research, education and action, 6(3)*, 381-387.

Schully, S.D., Benedicto, C.B., Gillanders, E.M., Wang, S.S., and Khoury, M.J. (2011). Translational Research in cancer genetics: the road less traveled. *Public Health Genomics, 14*(1), 1-8.

Van der Laan, A. L., & Boenink, M. (2015). Beyond bench and bedside: Disentangling the concept of Translational research. *Health Care Analysis, 23*(32), 32–49. doi:10.1007/s10728-012-0236-x PMID:23248053

Wendler, M. C., Kirkbride, G., Wade, K., & Ferrell, L. (2013). Translational Research: A concept analysis. *Research and Theory for Nursing Practice: an International Journal, 27*(3), 214–232. doi:10.1891/1541-6577.27.3.214 PMID:24422334

ENDNOTES

1 A similar approach should be taken to other issues in the R&D domain, such as measuring the efficacy of a TTU.

2 Deciding the allocation of public funds is best left for committees rather than single persons for litigation and appeal reasons.

3 The link was too long to be inserted in the text: http://www.idi.mineco.gob.es/portal/site/MICINN/menuitem.dbc68b34d11ccbd5d52ffeb801432ea0/?vgnextoid=b2a73c2b67c47510VgnVCM1000001d04140aRCRD.

4 Tax authorities usually prefer to have the investor prove the investment is lost before acknowledging the loss, then to do so at the time of investment off-setting it against other gains.

Chapter 6
Increased Complexity in Programs

ABSTRACT

The chapter discusses the issue of complexity in R&D support programs. The reasons for the complexity, the difference in the point of view of the government agency and the other players. With time and better understanding of the different needs of different sectors, different players – the programs tend to grow more complex. Due to different reasons such as: sectorial versus technological pressures, trying to do several things with one program, the need for different tools, the lack of other means trap – not all needs must be answered by a single program.

POTENTIAL PURPOSES FOR PROGRAMS

When we deal with the purpose of a program we need to define what we are referring to as a purpose. As explained above all the programs have the same final goal or purpose and that is the creation of an economic growth engine. However, while we have so far regarded R&D as the main tool of the program it is not the only available tool for the creation of economic growth.

On a more general scale, any innovation support program could potentially benefit the economic growth, as innovation is regarded as a major engine for economic growth. Thus, a program promoting the setting up of small businesses in a town, could be said to promote economic growth.

DOI: 10.4018/978-1-5225-3652-9.ch006

Example 1: Business Entrepreneurship Program

A program is offered where a partnership between a ministry in charge of economic growth and municipalities is set. The partnership sets up offices in specific towns offering coaching services for entrepreneurs as well as help in preparing a business plan and accompanying the entrepreneurs in dealing with the local banks to secure loans. The aim of the program is to help set up small businesses such as hairdressers, restaurants etc. and thus help the economic growth. The partnership subsidies the support given – but there is no R&D involved. The program is helped nationally by a central data base of experts, and financial support from the ministry with local offices and personnel paid for by each municipality.

While the above example is definitely not linked in any way to R&D and thus is clearly not of the family of programs that are the topic of this book, other programs are less clear cut.

Taking for another example the Danish "Industry PhD program" mentioned in the last chapter, which has been operating for a long period and while it operates via project finance of a specific R&D project, its main goal is to introduce PhD qualified personnel into the industry and thus improve its R&D infrastructure.This program as well as several of the Marie-Curie schemes operated under Horizon 2020 of the EU Commission, are aimed at improving the R&D personnel and its distribution. These programs while having stated goals of improving the existing R&D personnel, or introducing new personnel into industry and academy, changing the distribution of such qualified people etc., are in fact programs attempting to improve R&D performance in order to boost economic growth. It is important to note that most of them include some R&D project support and that for the following main reasons:

1. By supporting a project, the firm/research organization receiving the new person is deriving a direct benefit that it can evaluate and appreciate, and thus the entire program is viewed as practical.
2. The potential benefit of the employment of the person can be evaluated as this project serves as a test case for the benefit that specific person can bring in the future position he will hold. This is an important risk reducing measure as the cost of acquiring personnel, especially of a type unknown in the firm- where R&D activity has been low, or foreign, or directly from an academic organization with a different attitude, and organizational structure can be deemed costly and thus risky. Using

the program reduces the risk via the funding support, and thus, allows firms to feel more confident about the step.

3. It can be used to cause the recipient of the person to participate in the cost of acquiring the new person, and thus, present a measure of seriousness and commitment on the part of the receiving firm. This reduces the risk of free loaders for the supportive agency.
4. It provides a measurable support to the firm, one that can later be assessed, via the support to the firm, in addition to the inception of the person into the recipient. This is an important measure, as in cases where due to personal reasons the person does not remain in the recipient, it can still enjoy a benefit and perhaps try again learning from the past.

However, some programs, not referring to the last mentioned, have also a tendency to fall into the secondary goals trap, by focusing all the measurable effort on the number of people enrolled and supported, the number of people kept by the recipients after the expiration of the project and the support – the official goal of the program, and only rarely do these programs attempt to calculate the actual economic benefit to the firms or nation resulting from the activity. Granted, it is much more difficult to measure the last parameter, but after all, achieving the economic growth is the top-most goal of such programs.

One of the main pitfalls lies in the ability to measure the economic growth, but a higher hurdle seems to be getting the required information from the recipients. It is not just lack of cooperation on the part of the recipients, who may be reluctant to supply business information about their enterprises, but mainly their inability to isolate the specific contribution of the person to the firm from all the other contributing factors.

The programs therefore gravitate towards what is measurable and thus tend to focus on secondary goals.

It can be argued that if the program is designed correctly and according to the needs and barriers of economic growth, that the measurement of the secondary goals is good enough an indicator regarding the economic growth contribution. However, that statement does not take into account an unforeseen market failure, a value chain failure or any other break in the process linking the measurable result of the program and economic growth. If any of these breaks occur one might end up with a successful program achieving its measurable goals, but not contributing to the economic growth – and never know that.

There are no magic solutions that can be offered here. The need to develop both the measurable indices and regulations of reporting in the program before lunching it, is crucial. When the program is launched the reporting, obligations regarding it and its results should be clear. No participant should be able to refuse to submit such information based on lack of knowledge that it would be required, or lack of ability to support it. Firms especially should be made aware of the reporting requirements and should be prepared, including installed infrastructure to provide such information. This should not be seen as a deterrent for take up, as long as confidentiality is assured, and if the program is designed to answer a real need.

Other examples of related programs can be viewed as programs offering support for pre-R&D feasibility research, either business or technological. These are in many cases preludes to real R&D projects, which are intended to remove the less promising project proposal before a large financial commitment is made for their development. If you fail pre-R&D business feasibility testing, then this project should not pass the evaluation of the R&D proposal and the funds could be allocated elsewhere. Since entrepreneurs are optimistic by nature, these programs (e.g. SME instrument phase 1 in Horizon 2020) are presented as a hopeful aid for the next stage, but in fact they should be used to sift the less promising projects from the business point of view.

SINGLE PURPOSE PROGRAM

The simplest program aims at a single goal and is narrowly defined. As such it is normally simple to operate from several reasons:

1. It would fit the needs of a specific set of applicants, making the definition of the eligible for support applicant simpler and easier to enforce;
2. The support rules and the categories supported are narrowly defined towards the single goal, making it easier to operate the program;
3. Number of applications is normally limited due to the low number of applicants and that only a smaller number of applications from the total potential R&D applicants can apply for the specific goal.

However, such single purpose program will not last long as a unique program, as pressure would normally be applied to make sure that additional programs are initiated to answer additional needs, targeting complementing goals. It is important to recall that the programs we are discussing here are funded by public funds and that therefore there is a demand that the funds be utilized for the public good. The public good is more easily demonstrated when it applies to a larger number of people or applicant entities in our case. Therefore, a larger number of potential applicants, makes it easier to accept that the programs are for the public good.

Therefore, especially if successful, or deemed successful, the single purpose program will promote the initiation of additional programs. One point to notice is that it is not required that the program show final positive results at the earlier evaluations of the program, it is enough for the program to show that there is a big enough interest expressed as take-up or applications for the demand for additional programs to appear and increase. In the long term it is important to check for the final positive results.

COMPLEX: MULTIPURPOSE PROGRAMS

One of the dangers successful programs often face,is that instead of additional single purpose programs, the pressure for increased applicability of the program, may result in making the initial program more complex. That is to say, adding additional goals, and therefore complicating all of the above.

A multipurpose program needs first to define if the proposals should now address all the goals simultaneously or if addressing only some of them is enough. Demanding a complete compatibility with the goals / purposes would actually reduce the number of applicants and application, and therefore the take-up, which will result in a genitive reaction. Therefore, most such programs agree to accept also proposal with partial compatibility, or such that address only some of the goals. This, makes the management of the program more complex. When dealing with partial compatibility that allows different organizations, each addressing a different set of goals to apply for support for different projects, making the eligibility check more complex, making the evaluation and granting process more complex and open for criticism, and lastly making the management or accounting during the project and after the project end, more complex.

It is much simpler if the complexity is deferred to a set of single purpose programs that complement each other, but that can be managed separately, and therefore enjoy the benefits of the single purpose program combined with the needed coverage of the complex programs. That is to say that the single purpose programs cover the same set of needs but without the added complexity of the multipurpose programs.

One of the issues to discuss, is the ease of application. There are several factors complicating the application process:

1. The identification of the right program to apply to. That is especially true for cases when there are numerous programs, single purpose programs, and is even more problematic when the program names are not indicative of their nature, or goal.
2. The complexity of the goals makes the proposal more complex both to institute the eligibility of the applicant. This is especially evident when there is a demand for a consortium or partnership for applying to the program. Another aspect regarding this parameter are the preliminary demands on the applicant, regarding size, capabilities, past experience and financial viability.
3. The complexity of the goals is reflected as well in the work plan of the project. When more goals are presented there is a tendency, even if no such requirement is stated to comply with as many as possible, and this tends to complicate the work-plan.
4. Multiple goals in many cases require an increase in the number of supported categories. These categories when larger in number and therefore requiring more developed monitoring systems also require better understanding of the regulations, deciding which expanse should be apportioned to which category, both during the application phase as well as the execution phase. This two complicated the application, as the final goal of the application is the execution of the project successfully and not just applying for support.

These aspects and the resulted increased complication would act as deterrent, and actually reduce the number of applicants. Simpler programs allowing single applicant to apply for specially defined narrow goals of R&D projects would result in the largest number of applicants accumulated over all programs

EXAMPLE

When the EU Fifth Framework Programme for R&D was opened in 1998 it was a complex program (Cordis, 2009). It had four thematic programs offering to finance projects answering a certain theme and using as a major tool for that, an international R&D consortium. There were three minor supporting programs, e.g. support actions and human resource program which operated a limited number of projects aimed at some research mobility. Two rounds later the framework program was much more complex (European Commission, 2016). There were 10 thematic areas, operating several types of projects, including two variations on the international R&D consortium. There were several "horizontal", that is cutting across the thematic programs, support programs dealing with human resources mobility, and the framework has become the back office for other programs such as COST and the ERC. Horizon 2020 added some additional features to the mixture. While this may seem strange it is important to note that the financial support available has increased and thus also the responsibility to expand the target market for support. More tools to answer additional and more divergent needs. One should bear in mind the question, when does the equation of cost (learning to manage it on the market side) benefit (doing more good than before) change. Has the European R&D framework passed that point? Has its potential for effectiveness peaked?

IMPACT ON COMPLEXITY

The complexity of R&D support program is a result of several factors, which we shall discuss below, it may have inverse effect on the potential applicants. While it is easy to present the negative effect, increased complexity has on the take up of R&D support programs, it is not so easy to understand the reasons behind the complexity. In this section, we shall discuss the reasons for the increased complexity, and the impact it has on the applicants.

The Reasons for Complexity

The wrong assumption would be that there is no reason except laziness or lack of thought. As can be seen in the above-mentioned example of the evolution of the EU R&D framework programs, lack of thought or planning is not the case. One should not think that increasing the number of types of projects supported, or thematic programs is less cumbersome to plan than a simple program.

Anyone who went through any cycle of public funds allocation, and had to deal with the approval process and the demand for reasoning, explanations, potential impact etc. would know that lack of thought is not the issue. Faulty base assumptions - perhaps, lack of familiarity with the psychology of the applicants – probably. But the most prevalent reason is the under estimation of the impact complexity would have on the take-up, and over estimation of the power of the funding to drive the applicants to ignore the complexity.

Considering the example above, the officials designing the framework program for long periods before it is launched, have the time, and can afford to dedicate the effort to master the program on all its aspects and nuances. For even a medium size firm searching to promote one or even some of its R&D activities, this is not the case. Hence the rise in activity of different consulting entities who find that effort worthwhile.

Let us consider the specific reasons for the difference in views between the officials and the grant seekers:

Reason One: Over Estimation of the Power of the Funding

It is often considered in governmental circles that companies, being economic entities would go to quite extensive length to obtain funding, assuming that this is the most critical aspect of a firm's activity. While firms are economic entities and are very interested in funding, extending their resources and enhancing their value, many of them, would not do it at any cost.

For large firms that can easily be understood, they normally deal with funds several orders of magnitude larger than the potential funding from the R&D support programs, and could therefore, and in many cases actually they do, prefer to fund their own R&D without external funding, even when collaborating with external partners.

For small companies, especially high intensity R&D based firms, the most important aspect in many cases is not the financial report and direct funding, but rather the increase in the company value – increased IP ownership or access to it. For such firms for example, the potential of funding is much less important than the potential risk to their IP, or knowledge rights. Furthermore, these firms today have more and more available alternative funding, which is normally perceived as simple, and for their limited managerial capacity therefore much more interesting.

It is important to remember that for start-ups especially, the need to contend with a complex R&D support program, to learn the rules, apply, wait for the answer, make sure that all the issues in the application, award, execution, and post project periods are clear and not risky, can seem to be more intimidating than raising the same amount from their investors or new investors – which they have done and feel they know how to do. Not forgetting low success rate in obtaining the financial support in some of the programs (e.g. the Horizon 2020 SME instrument Phase 1).

A similar outlook exists for firms with little or no R&D. Here the lack of managerial experience and probably therefore of managerial capabilities to manage the R&D process and to derive benefit from it would act as a barrier towards applying for R&D support programs. The complexity would just add to barrier perception.

To summarize this point – there are large groups of potential applicants that the funding offered may not off-set the barriers to applications, perceived or real, and that in such cases the more complex the program the higher the barrier. Therefore, do not consider that complexity can be off-set by the funding offered by the program, it may have such effect, but in a limited way.

Reason Two: Easier to Manage and Fund

This is a cultural gap. Administrators consider that the administrative process clear to them, is also just as clear to people outside the administration. The process that may seem clear, simple and necessary to an administrator, may completely baffle a very wise engineer who can solve engineering problems the administrator could not comprehend. These are types of professionals looking at a process designed by one for the use of both.

Administrators in the governing entity may assume, quite often correctly, that it is easier to add supported activities and budgets to existing programs than approve new programs for these activities.

An active R&D program designed to assist firms in their own internal R&D, with a good track record and which both the firms and the governing agency may deem to be successful, would seem easier to treat as the basis and to add to it cooperation with the academy (and the relevant budget) than forming a new untried program. From the administrators who has to get the approval point of view, this is very logical. For the firms who now have to learn the new rules, and how to apply to the right track, and what to do when moving from one to the other, this may not seem as wise. Learning the new rules may goad potential applicants to prefer the older, known programs, thus becoming conservative in that regard.

So, while the administrators would assume that there is a need for the new activity, and that the firms would welcome the opportunity, the way it is done, may have a negative impact on both the new and the old activities and the way they are perceived.

There may well be additional influencing factors related to that reason, the limitation of allocating more resources (HR) for the new activity and therefore preferring to use the existing infrastructure. There may also in some cases, be legal barriers that make the option of adding to existing programs much more attractive than the independent alternative.

Summary of the Complexity

To summarize, while adding to existing programs may seem more efficient from administrative, organization, funding and similar reasons, and dealing with the added complexity may appear not to be such a big problem from the administrative point of view, in many cases that is not the same for the industry. The industrial point of view may be that the added complexity has made not only the new activities less attractive to apply for, but also the existing activities. Research organizations, much like the consulting firms, as they are in the business of submitting research proposals, and as they can afford it, may find it useful to invest in learning the rules. Which would make them less sensitive to increased complexity.

It should be remembered that the R&D programs are not an end but rather the means to promote economic growth. Allocating budget that could be spent elsewhere, to programs that have little or not take up in counterproductive and may lead to public criticism.

It is therefore important to develop the correct, albeit ever changing with time, balance regarding the complexity and simplicity of the programs, and to try and make other elements, more in the administrators' control simpler.

Recommendations for Simplicity

As we discussed above the number of types of expanses supported is limited. It would stand to reason that listing them with as little variation in reporting or in rate of support of each would make the budgeting effort in the proposal preparation phase, and the reporting during the execution phase much simpler. Employment expanses should fit the relevant sector (e.g. academic institutes have scholarships, firms rarely do), and the types bearing overhead should be relevant for the majority of projects expected – most would be based on

HR, but in cases of smaller firms where much may be outsourced, materials or the entire budget should also be considered.

Eligibility regulations should be simple and easy to demonstrate, requiring as little as possible external testimonials from the firm – legal adviser and accountant are Ok, more than that that (e.g. stock-owner declarations), and firms, especially SMEs would see this activity as a real barrier.

Regulations regarding the IP are crucial. The regulations must be clear and beforehand, and wherever there is a possibility of inequality in the ability to utilize or benefit from the research result among partners, it is recommended to prepare, and sign an agreement in that regard, settling any potential disputes. The program should also define and present beforehand a clear mechanism for dispute settling regarding IP issues, preferably with the governing agency as an oversight. It is important to make the rules clear regarding both ownership and utilization, so the applicants know what they are getting into beforehand.

Continuance, reimbursement and complaints are issues that should also be described in a simple and clear way.

REFERENCES

Cordis. (2009, July 16). Fifth Framework Programme. Retrieved from http://cordis.europa.eu/fp5/about.htm

European Commission. (2016, January 25). Research and Innovation: FP7. Retrieved from https://ec.europa.eu/research/fp7/index_en.cfm

European Commission. (n.d.). Marie Skłodowska-Curie Actions-- Horizon 2020. Retrieved from https://ec.europa.eu/programmes/horizon2020/en/h2020-section/marie-sklodowska-curie-actions

FondenI. (n.d.). Industrial PhD. Retrieved from https://innovationsfonden.dk/en/application/erhvervsphd

Chapter 7
Research Policy and Adjacent Tools

ABSTRACT

In this chapter, we discuss the elements required to complete the picture. The issue of Research Policy, its role and impact is presented. Following that discussion are the presentations of additional tools to the R&D Support Programs support tools – Accelerators, Technology Incubators and more. The presentations end with the National Research Organization and the comparison of such organizations with the R&D support programs.

INTRODUCTION

When we discuss Research Policy we need to ask first, why is such a policy required? Since this is a wide topic and not the main focus of our discussion, yet too important not to mention it, we shall answer the question from the practical aspect.

Having a policy, discussed and approved, with allocated resources, allows for improved efficiency in future allocation of resources, in promoting required legislation and in blocking unwanted changes that could have a negative impact on the policy.

Even at the outset, as deciding and approving a policy is a lengthy process requiring substantiation and proof of concept, just that planning and thinking effort is a benefit for most countries.

DOI: 10.4018/978-1-5225-3652-9.ch007

However, all the qualities above make having a policy a liability of sorts. It means that by having a policy one trades off efficiency for flexibility. Flexibility or rather ability to reallocate resources in our case for topics or actions that were not foreseen, or not selected as part of the policy is lengthy, problematic and in some cases not feasible within a relevant frame of time. That reduced flexibility is not a minor problem. If the world is changing in ever increasing pace, the world of knowledge is certainly doing that. The advances require more effort on the R&D front, to keep up with changes and challenges and that requires some degree of flexibility which to a certain degree does not agree with an efficient policy.

However, we are not going to continue the discussion about the efficiency versus flexibility, but rather consider the implications of having a policy. In places where policy does not exist, it is replaced with short term or ad-hoc decisions.

Our interest in research policy stems from having research policy as the basis for R&D support programs and their resources.

SETTING RESEARCH PRIORITIES

Setting research priorities in our discussion refers not to the process of selecting these priorities but rather to the implications of having such priorities selected within a policy (Choi and Lee, 2000; Schmid, 2012; Shao and Yuan, 2013; Maleki et al., 2014).

It is important to note that while selecting the priorities for the research policy one does not only selects what will receive assistance, but rather, in most cases, what will not receive such support.

When setting priorities to support strong industrial sectors, or a strong academic heritage, such a policy will further oppress other sectors and can perhaps lead to their demise in the region, or country. If for examples, from political reasons atomic energy research is removed from the list of supported technologies or field of research, one would see first the firms involved in the sector and later the academics leave. They will find other countries where such topics are supported, and quickly enough the entire sector and the relevant disciplines will disappear.

It is especially easy for academics to migrate, as they do not physically need to emigrate but can only change, and that is difficult enough, their research topics. Such a move would require establishing new interest, creating a new list of publication and alternative reputation. In most cases one would see some of the academics physically emigrate and some change their research interest. For firms, unless they sell only to the disappearing local market, emigration to other countries is the most common solution. Changing the market, they are active in, is in some cases even more difficult. However, if they want to keep to their research topic physical migration is often easy as they would have an international reputation to support that move.

Of course, not all policies need be so harsh, many can offer some reduction in the support, which is critical by itself but may be less dramatic.

The topics selected for priority therefore should be selected with care. In some places, the emphasis is placed not on sectors with a strong lobby, but rather on weak or small sectors which for political reasons are selected for support and development. Here the potential risk is double: there may not be enough firms for developing the sector and thus while other sectors are hungry for assistance, the selected sectors are having more funds than they can use, or that support is being diverted through loopholes into other, not referred sectors.

One of the major traps is trying to select the main topics for future technological market disruptions, to predict the unforeseen breaks. There is a logic trap in such an effort and in the end, it requires a wide enough band of selected topics for support in order to do some risk management.

It is recommended, if it is possible, to neutralize political influence to try and select a combination of topics that it is deemed important to support and promote, but to create the specific niche programs for the topics that one understand it is important to preserve. In the example mentioned above, a program dedicated to nuclear research, involving academia and industry, but with a limited set of funds, and with a demand that each proposal present access to international markets, could help maintain whatever industry exists and even the professional training expertise in the country.

SETTING PROCESS PRIORITIES

The topics are important but just as much are the types of projects and the process.

Selecting a topic will focus attention on related research, but selecting the process under which it will be supported and the type of research supported – the

R&D process to be supported are critical, as they determine the participant's type and thus the wellbeing of the other types.

Selecting a topic such as, renewable energies is interesting and can open many possibilities. However, selecting a program that allows firms to apply for development projects, academy researchers to apply for basic research, or consortia of both academy and industry for the development of basic technologies, will have a totally different impact than the decision to support the topic via the creation of the national research institute for renewable energies.

It is important to understand such a decision on a strategic basis. While such an institute will require longer in setting up the infrastructure (machine as well as HR), and getting up to speed, once it is fully operational it will be very efficient in supporting whatever industry there is. One should also remember that shutting down such an institute is not easy, nor is it fast and short.

The different programs, and by the way, selection of several is also an option, are better suited for a quick change. If the players are in place, the work, the projects can start instantly and the results can be seen fast enough, it is also easier to change. It also means that direct support is given instantly to the existing infrastructure, furthering its development both in industry and academy and encouraging more activity in the field. It shall be less efficient in the future as the finding of partners is more difficult than direct access to a known research organization, but that is in the long run in most cases.

Therefore, when making a policy decision, one should recall that the tools, the processes to be employed in the execution of that policy will have long and short-term impact on the policy enactment results and that therefore it is important to include these considerations when making the policy decision.

THE IMPACT OF RESEARCH POLICY

Research policy can be designed for several reasons. The most common is to support the economic growth in the country by supplying support to R&D activities in the hope that it will help the industry via increased innovation to increase its competitiveness and thus improve its revenues (Chan and Daim, 2011; Moon and Cho, 2014; Flipse, van der Sanden and Osseweijer, 2014).

However, deciding to support traditional and rural areas only, in order to prevent brain, drain, or the emigration of young people to the cities, or from other demographic reasons, is also a common enough aspect in setting research priorities that support such activities. Another can be the wish to preserve certain sectors from political reasons, such as agriculture, forestry and other industries that are important nationally and that require support and that the R&D support is deemed as a more legitimate subsidy then direct funding or other solutions.

When Germany decided in the early 2000's to support the development and installation of solar based sites, it allowed the creation of a PV sector as well as other solar technologies in Germany, that would otherwise have never developed to such an extent, based on local demand alone. Understanding that in the infrastructure markets, the main entry barrier is performance statistics and operating sites and unlike in communications - the latest technology, German government gave the local firms the ability to enter world markets with significant presence, which is translated into Income, training and employment, and to a certain extent creates a strong sector where a much weaker one would have otherwise developed. Had Germany decided to support only local academic research into renewable energies, the result would have been much less impressive. Therefore, when setting the research priorities, and when setting a research policy, one should remember the potential impact the policy will have and should design the policy accordingly. The correct support is a question of the topic selection, the tools and process for execution and any specific requirements that may help reach the wished-for impact.

ADJACENT TOOLS

The adjacent tools are the additional organizations or formats of operation that have sprung into existence in order to facilitate part or the whole of the knowledge transfer from creators to market utilizers.

These cannot be viewed as externals to the innovation or research policy as they come to serve it and promote its aims and goals. However, they are not R&D support programs, and in most cases, are not facilitators but participants in the R&D process. Further to that, while it is not compulsory, they may be applicants to some of the programs mentioned above as per their role in the R&D process. Some of the main tools and especially the main functions that they come to fill are mentioned below. It is evident that the dedication of a whole chapter cannot discuss each of the tools to the full, and of course not a sub-chapter dedicated to each. However, within the scope of this book this discussion fills the role of presenting the main aspects relevant to the adjacent tools. In some of the cases presented below, we present also an R&D support program aiming to fill the same or similar role.

Accelerators

The main role of accelerators is to present the technologies or ideas to be developed to the first investment round and the first tier of early investors (referred to in most cases as "Angels"). Investments are mostly dedicated to a preliminary feasibility study, mostly non-technological, but not excluding technological proof of concept (POC), as the sums invested are not enough to allow for a full-scale feasibility study, but are focused on a very preliminary technical feasibility and mostly on the business side as well as the legal / IP issues relevant for the future project. As it is a very early stage in the formation of a future company towards the introduction of the technology and resulting products to the market, its main role is to transfer the nascent firms to the next stage at the best possible condition for further investment and further development.

The accelerators develop close to larger organizations that can either foster the technology, for the future companies or are interested in developing the economic activity of new companies, and attracting investments and angels to the region. Therefore, one can find accelerators on academic campuses, industrial campuses supported by large firms owning the campus, owned and operated by municipalities and similar organizations.

The duration of the stay of the projects in the accelerators is normally short, in some cases limited to six months – year of work; assuming that failure to mature to the next stage by that time means that the project has to be scratched, as both the time and funding should by this time, have resulted by strong enough results. Failure to do so reduces the ability to raise further funds (assuming that the original investment has already been utilized), and that therefore the project is dead.

The management of the accelerator rarely looks for revenues from the rent and support services as the main source of funding, but would seek some compensation for the services, as well as some stake in the future firm, when and if it would have value one day. The dependence on large organizations, either large firms, universities municipalities etc., is a result both of the risk involved in supporting such high risk early projects, as well as the time required for the successful ones to mature towards a stage where the investment can be cashed.

A parallel to the accelerators in the R&D support programs one may find the Horizon 2020 SME instrument program first phase which also supplies

50,000 euros for 6 months for a business and IP feasibility study (full details on the Horizon 2020 website).

Technology Incubators and Technology Parks

Technology incubators and technology parks mean different things in different countries. The most supportive structure of a technology incubator is the one which supplies both the support environment for the development of the new firms' technologies as well as the funding for their activity in return for some equity. The funding may come from a combination of private funds as well as government support, or by completely private investment. When the funds are also from government sources, there are normally additional regulations and requirements for receiving that support, while the private funds normally entail only an investment agreement.

Some technology incubators provide the facilities, and charge for general services and rental fees. In many cases by providing also access to other facilities they combine their fee with equity holdings. In some countries, technology incubators accept firms in a more advanced stage than accelerators, and are therefore dealing with more mature firms, past their preliminary feasibility proof projects, and into a more advanced technology feasibility proof projects. The duration of the stay in such incubators is normally two to three years, and the size of investment in normally larger, enough to allow for the technology feasibility projects to be executed.

The advantage of the technology incubator[1], is that it offers services, especially management services, in reduced prices for the firms resident in it, and thus allow them to get the services they need at a price tailored for their need. The aspect of sharing the residence with similar firms, and the centralization of marking and fund raising together, give the activity added value. One must bear in mind that at a certain stage, the firm will have to leave the technology incubator, and to develop the capabilities for the services it has received so far from the incubator.

To a certain extent, the parallel for the technology incubator is an R&D support program, that supports the development of technologies, at least up to prototypes, but here the parallel is a little less clear, as the incubator supplies additional support in services. There are programs aimed at the support of new companies in their early stages, but they would in most cases propose a shorter duration. The additional services, such as different consulting services and mentoring which the incubator can offer, are normally not included in R&D support programs (The EU Horizon 2020 SME Instrument third phase being the exception rather than the rule).

Further to that, the incubators naturally give preference to technology commercialization by spin-offs, versus licensing to existing companies, which is of course legitimate, but has its own downsides.

In that regard, technology incubators may, over time, or by strategic decision from the start, develop an affinity to a certain sector (e.g. environmental technologies, medical equipment), and develop additional services in support of that affinity[2], as well as cultivating post incubators investors from the relevant sector, or with strategic interest in that specific sector. These incubators will naturally prefer project companies that would require these services, and would be interesting for the post incubator investors, in order to justify their interest in the incubator. This will create a positive feedback cycle, enhancing the validity of the services offered, as well as attracting the relevant investors.

Venture Capital Funds/Start-Up Stock Exchanges

The VCs and the start-up stock exchange are presented together although being different tools, as they serve to a certain extent as similar tools for the financial support of start-ups.

The VCs present an investment tool for a cycle of between 5-7 years in which a new company is supposed to present a significant increase in value. The VCs are not long-term investors that will accompany the firm all the way to maturity, unless this happens very fast. The VCs are there to invest the funds in high-risk ventures and to get out, hopefully with a good return on their investment within the set period mentioned above.

There is a lot to say regarding the activities of VCs, their impact on economic growth and the role they play in shaping the high-tech sector. However, for the purpose of the discussion in this book and the relevance to R&D Support programs it is enough to regard them as financial tools for start-up firms, supporting the early stages of the firms' lives.

It is important to recall that the investment or VC finance is normally between 0.5 million $ to several millions, and that the VCs tend to invest in projects with a clear and evident exit – recuperating the investment. The VC does that in exchange to equity in the firms, which it hopes to monetize at the exit to present profit on the investment.

The VCs differ from regular commercial banks as their type of investment differs significantly from the type of investment commercial banks are used to. Most of the start-ups do not have any collateral or income and therefore cannot show guarantees or a regular flow of income to return a commercial bank loan. Moreover, the value that the star-ups will create cannot be measured in the usual way in which these banks use when assessing assets. Therefore, while banks, where allowed by law may invest in the VC, they do not engage directly in venture financing unless it is according to their regular way.

The start-up stock exchange (see https://www.startup.sx) differs than the VC's in the fact that it is a market for these special firms. A market with expert vetting of the firms seeking to be traded in it, and thus more fitting to the characteristics and needs of a startup. The quality depends on the vetting and value setting. The investment is made by investors in the stock exchange much like other stock exchanges and is therefore spreading the risk among numerous investors, allowing for riskier investments. The stock exchange also enjoys the benefit that the investors investing in it already have an exit way – just sell your stock on the stock exchange, or if the firm is overly successful and is now traded in other places – trade it there.

Therefore, the start-up stock exchange is similar to VC in the investment in small high-risk start-ups, but differs in spreading the risk among numerous investors and in offering an exit for the investors. There are, of course, also platforms for crowed sourcing such investments (see https://crowdcube.com) and other financial support tools.

Tax Incentives

Tax incentives are another financial tool. In this case, one operated by the government and intended in most cases not for start-ups but for mature firms in the market – those that have a taxable income[3]. The purpose of the tax incentives in to encourage firms to invest in R&D activities and can be viewed as complementary to the R&D Support programs, as this is in a way also an investment of public funds in R&D. The concept is that firms may not like to expose their plans before the project. Further to that they would not like to expose their success and failures in R&D to the government after the project as well. However, these firms would still want to enjoy support similar to the one given to firms applying for grants from the R&D support programs. This can be achieved by getting tax exemptions or tax reduction proportional to their R&D efforts. This way if the firm does not want to disclose the attempts or results of a research effort it has already made, it can "pay the price of silence" and not submit any information. However, if at the end, it realizes that it would prefer to see some return for the effort, it may submit the request. It allows the firms better control over the issue of publication versus remuneration. This is different than the regular expanses report, in the fact that in most cases a detailed description of the effort made, the research plan and other detailed is required.

It is a supplementary tool, since firms applying for a tax incentive cannot be at the same time supported by an R&D support program – double reimbursement. Another limitation is that the incentives are given to the applicants only and not to their partners, unless they apply as well. It is also supplemental to the VC's and Start-Up stock exchange as it addresses mature firms active in the market. It also differs among the countries, as per their taxation system.

National Research Organizations

The debate about the role on national or regional research organizations is still on-going. There are several models, dependent upon the national legal system, the activities and nature of the academy and the size and variation in the national industry.

In countries like Israel, where the universities play the role of knowledge suppliers, and cooperate directly with the industry under national programs or privately, the national research organizations have a smaller role and not many exist. In other countries such as France, Germany, the Netherlands and Italy, where the gap between the academy and industry is bigger, where strong industrial sectors are very diversified and requires specialization, such organizations exist, sometimes in whole networks and play an important role in the transfer of the technology from the universities and other research organizations (e.g. in Germany Max Plank Institutes) to the industry. The research the national research organizations perform, is in many cases translational research, translating basic curiosity driven research of the academy into applied research to be taken on by the industry.

National or regional research organizations in their traditional form comprised of an infrastructure of buildings and research equipment, combined with HR infrastructure of researchers and technicians highly trained to operate the research equipment and to perform the required translational research.

The alternative of a virtual institute is in many aspects so similar to an R&D support program that we shall not discuss it here separately.

The establishment of a traditional national research organization therefore would require a significant investment in buildings, equipment, and personnel to work the equipment. It would further require support for the first period of operation to allow it to generate enough income to become independent. Since these are national or public organizations, the personnel in many cases is considered as public service and that with the sunk cost of initial investment makes the institute less flexible in employment and activity.

The above states, that there is in many such cases a tradeoff between flexibility, very often evident in the R&D support programs, regarding the sector, technologies and type of activity supported, and high efficiency on the other hand, accumulated experience in one center – as is evident in the national research organizations mentioned above..

An analysis similar to the one performed here regarding the R&D support programs (and which we shall not go into in this book) can be applied to the National research organizations, to analyze their activities, structure and impact.

REFERENCES

Chan, L., & Daim, T. U. (2011). Technology transfer in China: Literature review and policy implications. *Journal of Science and Technology Policy in China*, *2*(2), 122–145. doi:10.1108/17585521111155192

Choi, Y., & Lee, J. J. (2000). Success factors for transferring technology to spin-off applications: The case of the technology property rights concession in Korea. *The Journal of Technology Transfer*, *25*(2), 237–246. doi:10.1023/A:1007889010655

Flipse, S. M., van der Sanden, M. C. A., & Osseweijer, P. (2014). Setting up spaces for collaboration in industry between researchers from the natural and social sciences. *Science and Engineering Ethics*, *20*(1), 7–22. doi:10.1007/s11948-013-9434-7 PMID:23467918

Israel Innovation Authority. (n. d.). Innovation Israel. Retrieved from http://innovation-israel.mag.calltext.co.il/magazine/45/?article=4

Maleki, K., Hamadeh, R. R., Gholami, J., Hamid, S., Butt, Z. A., Bin Saeed, A., & Majdzadeh, R. et al. (2014). The Knowledge translation status in selected Eastern-Mediterranean universities and research institutes. *PLoS ONE*, *9*(9), 1–11. doi:10.1371/journal.pone.0103732 PMID:25197834

Moon, S., & Cho, S. B. (2014). Differential impact of science policy on subfields of human embryonic stem cell research. *PLoS ONE*, *9*(4), 1–7. doi:10.1371/journal.pone.0086395 PMID:24717403

Schmid, G. (2012). Technology transfer in the CDM: The role of host-country characteristics. *Climate Policy*, *12*(6), 722–740. doi:10.1080/14693062.2012.675733

Shao, Z. L., & Yuan, F. (2013). Research and design on national basic research management information data center. *Applied Mechanics and Materials*, *321-324*, 2602–2606. doi:10.4028/www.scientific.net/AMM.321-324.2602

ADDITIONAL READING

De Paula Fonseca e Fonseca, B., & Zicker, F. (2016). Dengue research networks: Building evidence for policy and planning in Brazil. *Health Research Policy and Systems*, *14*(80), 1–10. PMID:27825383

Kremic, T. (2003). Technology transfer: A contextual approach. *The Journal of Technology Transfer*, *28*(2), 149–158. doi:10.1023/A:1022942532139

Kumar, U., Kumar, V., Dutta, S., & Fantazy, K. (2007). State sponsored large scale technology transfer projects in a developing country context. *The Journal of Technology Transfer*, *32*(6), 629–644. doi:10.1007/s10961-006-8880-7

Luo, J., Flynn, J. M., Solnick, R. E., Ecklund, E. H., & Matthews, K. R. W. (2011). International stem cell collaboration: How disparate policies between the United States and the United Kingdom impact research. *PLoS ONE*, *6*(3), 1–7. doi:10.1371/journal.pone.0017684 PMID:21408134

Matthieu, M. M., Bellamy, J. L., Pena, J. B., & Scott, L. D. Jr. (2008). Accelerating research productivity in social work programs: Perspectives on NIHs postdoctoral T32 research training mechanism. *Social Work Research*, *32*(4), 242248. doi:10.1093/swr/32.4.242 PMID:28316462

Mujumdar, P. P. (2015). Priorities for sciences in India. *Nature*, *521*, 151–155. doi:10.1038/521151a PMID:25971494

Seeck, H., & Laakso, A. (2010). The adoption of management paradigms in Finnish management research 19372007. *Journal of Management History*, *16*(2), 174–194. doi:10.1108/17511341011030093

Shuyuan, X., & Huijun, M. (2016). Analysis on the impact of international technology transfer on China's industrial restructuring and upgrading. *International Journal of Innovation, Management, and Technology*, *7*(3), 86–89.

ENDNOTES

1 Pure business incubators would not be classified in most cases under R&D support programs.

2 In medical devices for example the introduction of an FDA approval consultant into the incubator offered services – would encourage the incubator to introduce projects that would pay for that service.

3 Except when the tax incentives are for the investors in the startup.

Conclusion

FUTURE TRENDS

The future trends of national R&D programs seem to be problematic. The increased demand to innovate pushes more organizations that have never done R&D before into that activity. On the one hand, the understanding that the there is a great diversity in the R&D support needs of different entities is increasing and therefore at the same time the diversity of the programs involved or created is increasing as well. However, this plethora of possibilities is making it even more difficult for potential user of such programs to identify the specific program best suited for their needs.

See for example the EU Framework Program (Horizon 2020) which has grown from five simple thematic programs in FP5, into many more different programs and sub-programs, all aimed in supporting R&D and related activities. Consider the complexity of the instructions a single program would have. Than consider such a big number of different programs, not just thematically, but also regarding the process of submission, type of activity supported and type of applicant (single person, legal entity, consortium etc.). This forms a barrier and deterrent to future applicants.

This situation leads to several phenomena that can be observed. The first is the emergent importance of the consultant, whose main expertise is the identification of the most suitable program to apply to. This is not a negligible issue as this raises the cost of submission and thus the barrier for entry, and therefore the take-up of the program, for some of the potential applicants. The second, is the tendency of applicants to become very conservative and to focus their efforts on the programs they know, even at the cost of missing a more suitable program to their needs. That second occurrence stems from the insecurity the potential applicant feels when such a wide variety of possibilities is encountered.

Another aspect to be considered is the fact that the available funds for the support do not necessarily increase in proportion to the increase in the number of programs, which results in effectively lower available funds for support. If you combine the last two aspects, the reduction in available funds, and the conservatism of the applicants, preferring to focus on the familiar older programs, it would mean that the competition for the funds grows every year, and the firms feel as if funds are scarce while officials note that this is not the case.

High competition on limited funds, would result in low success rates (in the Horizon 2020 some programs publish single digit percentage success rates), which in turn will create a pressure to reduce the number of programs, increasing the available funds in the main stream programs. This trends with a growing pressure for simplification will grow as long as current trends continue.

Another on-going trend that can be seen, not necessarily in Horizon 2020, is the increase and action taken to increase the available funding for activities that seem to have political priority. Funds for SMEs, or the diversion of such funds to the private sector is a growing trend which is noted and complained about in the academic circles as the funds for them do not increase at the same rate. The emphasis on the utilization of research results and promoting them towards the market, can have two aspects: one is to invest in increasing the rate of transfer to the market, and the second is to increase the funds available to applied research instead of basic research – it is this possibility that worries the long-term planners in the academic world.

SUMMARY

The role of R&D support program in economic growth is increasing. With the advent of the perception that innovation is the key element in maintain and developing a competitive advantage in the market, the different sources of innovation are receiving ever increasing attention.

One of the major sources of innovation, although not the only one, is R&D. Therefore, the support given to the R&D activity via programs or through the establishment of entities dealing with that activity as a major activity for them, is gaining increasing interest in understanding the mechanisms, involved entities and their roles and interests, as well as other parameters that lead to success or failure in this endeavor.

While the R&D support program has been active in most places since the 1980's and even before, it has never before gained so much interest. Furthermore, with the increasing global competitiveness and the announcement of knowledge as the key to economic success additional public funds have been allocated to that activity under different frameworks, and as such, public interest to understand the activity and the justification for investing so much of its funds in it, has also been growing.

In order to make some sense of the different players in different countries, under different legal systems and regulations, and sometime with similar names denoting different entities, this book has tried to paint a clear and concise picture.

The book is therefore listing the main and secondary participants, their interests and needs, the different points of view, the different organizational cultures and the difficulty all this plethora of parameters, bring into the attempt to understand the R&D support program. It also attempts to discuss some of the elements involved – types of research and IP issues as well as the different types of programs and the stages of R&D.

It is quite clear, that most of the topics listed above, are worth their own volumes and in some cases certainly many are available. However, here a comprehensive, even if not a complete enough picture is painted.

The reader should avoid the mistake of thinking that R&D is the sole source of innovation, and that the combination of R&D programs is the whole innovation system in a definite geographic area. Hopefully that is not the case, other innovation activities such as VC, Inland Direct investment, non-R&D innovation activity such as business entrepreneurships etc., should also be available in order to complement the R&D activity. It is especially important for entities that are not technological by nature, such as government and services, as well as commerce who are non-the less important for economic development.

Innovation being one of the most complex topics, even for entrepreneurs in the issue of knowledge management, is wider in scope than just knowledge protection. The future, if it is indeed based on knowledge, would require more and more experts in that area, and even low tier management would be required to be more versed in that topics in the future.

Similarly, the cooperation between the different entities regarding R&D activity will increase, both regarding the transfer of knowledge along the research chain as well as horizontally among different disciplines, and that will require better understanding of the barriers towards such cooperation.

The description of the origins of such barriers as well as ways to overcome them, and the role the government can play in that, are described and carry significance for future endeavors to develop the cooperation.

One of the major issues, still open for discussion, is the preference of the R&D support program, versus the establishment of the national research institute. Each alternative has its own advantages, and disadvantages, and even some combined models have evolved. This debate would go on for years to come as political as well as professional arguments enter the discussion. Here the issue has been presented but no final recommendation has been offered.

One of the major factors that has not been adequately discussed, due to the need to focus on the basics, and that is the issue of time. The booked touched the subject, but the time element in developing the abilities to cooperate as well as the road along the research chain, are a major factor in understanding the process.

There are other topics as well, that could not be fully discussed in such a book without losing focus. The role and function of the different interfaces such as the regional or national research center and the TTU (Leute, 2005; Marin et Al, 2015) have been mentioned, but much more could be said about them, and the functions that they serve.

In that regard one more point should be raised – the personnel in the TTU is in most cases incidental (Friedman, and Silberman, 2003). Not many have started with that specific career in mind (Dvoyris, 2017) and more effort should be made in encouraging activities like Pro-Ton Europe, in the support of developing the TTU business developer, as a profession. Especially in view of the role of the knowledge in the future economy and the Technology Transfer in the R&D process.

REFERENCES

Dvoyris, V. (2017). University Venture: The View From Stanford. *Coller Venture Review, 4*. Retrieved from https://www.collerinstituteofventure.org/university-venture-the-view-from-stanford/

Friedman, J., & Silberman, J. (2003). University technology transfer: Do incentives, management and location matter? *The Journal of Technology Transfer*, *28*(1), 17–30. doi:10.1023/A:1021674618658

Leute, K. (2005). Patenting and licensing of university -based genetic inventions- A view from experience at Stanford university office of Technology Licensing. *Community Genetics*, *8*, 217–222. PMID:16244475

Marin, A., Boanta, L., Hadar, A., Badea, D. M., Vladut, G., Bucur, D., & Ivan, I. et al. (2015). Business models and competitive advantage for technology transfer entities. *The Romanian Review of Precision Mechanics*, *48*, 103–109.

Related Readings

To continue IGI Global's long-standing tradition of advancing innovation through emerging research, please find below a compiled list of recommended IGI Global book chapters and journal articles in the areas of research and development, research methodology, and business research. These related readings will provide additional information and guidance to further enrich your knowledge and assist you with your own research.

Ahmed, B. S., Ben Maati, M. L., & Al Mohajir, B. (2015). Improve Intelligence of E-CRM Applications and Customer Behavior in Online Shopping. *International Journal of Business Intelligence Research, 6*(1), 1–10. doi:10.4018/IJBIR.2015010101

Anagreh, L. F., & Abu-Shanab, E. A. (2017). Voters Intention to Use Electronic Voting Systems. *International Journal of E-Business Research, 13*(3), 20–38. doi:10.4018/IJEBR.2017070105

Antunes, A. M., Mendes, F. M., Schumacher, S. D., Quoniam, L., & Lima de Magalhães, J. (2014). The Contribution of Information Science through Intellectual Property to Innovation in the Brazilian Health Sector. In G. Jamil, A. Malheiro, & F. Ribeiro (Eds.), *Rethinking the Conceptual Base for New Practical Applications in Information Value and Quality* (pp. 83–115). Hershey, PA: IGI Global. doi:10.4018/978-1-4666-4562-2.ch005

Ardyan, E. (2015). Market Sensing Capability, Entrepreneurial Orientation, Product Innovativeness Success, Speed to Market and SMEs Performance. *International Journal of Business Intelligence Research, 6*(2), 18–32. doi:10.4018/IJBIR.2015070102

Ardyan, E., & Aryanto, V. D. (2017). The Importance of Electronics News Brand Trust: The Case of Online Newspapers in Indonesia. In I. Oncioiu (Ed.), *Driving Innovation and Business Success in the Digital Economy* (pp. 99–118). Hershey, PA: IGI Global. doi:10.4018/978-1-5225-1779-5.ch008

Baporikar, N. (2017). Innovation Implementation: The Critical Facets. In I. Oncioiu (Ed.), *Driving Innovation and Business Success in the Digital Economy* (pp. 41–54). Hershey, PA: IGI Global. doi:10.4018/978-1-5225-1779-5.ch004

Baranowska-Prokop, E., & Sikora, T. (2017). Competitiveness of Polish International New Ventures from Managerial Perspective. In A. Vlachvei, O. Notta, K. Karantininis, & N. Tsounis (Eds.), *Factors Affecting Firm Competitiveness and Performance in the Modern Business World* (pp. 83–107). Hershey, PA: IGI Global. doi:10.4018/978-1-5225-0843-4.ch003

Barnett, R. (2017). Researching Supercomplexity: Planes, Possibilities, Poetry. In L. Ling & P. Ling (Eds.), *Methods and Paradigms in Education Research* (pp. 291–308). Hershey, PA: IGI Global. doi:10.4018/978-1-5225-1738-2.ch018

Bazargani, M., & Namazi, E. (2015). A Study on a Combined Model in Business Intelligence for Improving Electronic Insurance. *International Journal of Business Intelligence Research*, *6*(1), 49–55. doi:10.4018/IJBIR.2015010104

Behboudi, M., & Mokhtari, H. (2017). Online Advertising: Experimental Facts on Ethics, Involvement, and Product Type. In I. Oncioiu (Ed.), *Driving Innovation and Business Success in the Digital Economy* (pp. 119–136). Hershey, PA: IGI Global. doi:10.4018/978-1-5225-1779-5.ch009

Besana, A. (2017). Animal Spirits or Engaging Spirits?: The Importance of the Not-for-Profit Economy – Coopetitiveness and Grant-Making Foundations. In A. Vlachvei, O. Notta, K. Karantininis, & N. Tsounis (Eds.), *Factors Affecting Firm Competitiveness and Performance in the Modern Business World* (pp. 223–250). Hershey, PA: IGI Global. doi:10.4018/978-1-5225-0843-4.ch008

Bird, J. L., & Wanner, E. T. (2015). Research as Curriculum Inquiry: Helping College Students with Anxiety. In V. Wang (Ed.), *Handbook of Research on Scholarly Publishing and Research Methods* (pp. 272–294). Hershey, PA: IGI Global. doi:10.4018/978-1-4666-7409-7.ch014

Boeren, E. (2015). Surveys as Tools to Measure Qualitative and Quantitative Data. In V. Wang (Ed.), *Handbook of Research on Scholarly Publishing and Research Methods* (pp. 415–434). Hershey, PA: IGI Global. doi:10.4018/978-1-4666-7409-7.ch021

Boukhedouma, S., Alimazighi, Z., & Oussalah, M. (2017). Adaptation and Evolution Frameworks for Service Based Inter-Organizational Workflows. *International Journal of E-Business Research*, *13*(2), 28–57. doi:10.4018/IJEBR.2017040103

Brewer, E. W., Torrisi-Steele, G., & Wang, V. C. (2015). Survey Research: Methods, Issues, and the Future. In V. Wang (Ed.), *Handbook of Research on Scholarly Publishing and Research Methods* (pp. 396–414). Hershey, PA: IGI Global. doi:10.4018/978-1-4666-7409-7.ch020

Briones-Peñalver, A., & Poças Rascão, J. (2014). Information Technologies (ICT), Network Organizations, and Information Systems for Business Cooperation: A Focus on Organization and Strategic Knowledge Management. In G. Jamil, A. Malheiro, & F. Ribeiro (Eds.), *Rethinking the Conceptual Base for New Practical Applications in Information Value and Quality* (pp. 324–348). Hershey, PA: IGI Global. doi:10.4018/978-1-4666-4562-2.ch015

Brookfield, S. (2015). Preparing Book Proposals for Scholarly Publishers. In V. Wang (Ed.), *Handbook of Research on Scholarly Publishing and Research Methods* (pp. 1–20). Hershey, PA: IGI Global. doi:10.4018/978-1-4666-7409-7.ch001

Buttram, J. L., Archbald, D., & Farley-Ripple, E. N. (2016). Addressing Critical Problems through Leadership Portfolios: A Content Analysis. In V. Storey, & K. Hesbol (Eds.), *Contemporary Approaches to Dissertation Development and Research Methods* (pp. 238-254). Hershey, PA: IGI Global. doi:10.4018/978-1-5225-0445-0.ch015

Calway, B. (2017). Work-Integrated Learning Praxis: Selecting a Research Paradigm. In L. Ling & P. Ling (Eds.), *Methods and Paradigms in Education Research* (pp. 278–289). Hershey, PA: IGI Global. doi:10.4018/978-1-5225-1738-2.ch017

Cambiano, R. L., Speaks, P. C., & Cambiano, R. M. (2015). Infusing Yourself into the Backstory: A Multidimensional Case Study Perspective. In V. Wang (Ed.), *Handbook of Research on Scholarly Publishing and Research Methods* (pp. 349–374). Hershey, PA: IGI Global. doi:10.4018/978-1-4666-7409-7.ch018

Cantatore, F. (2017). The Migration of the Book across Territorial Borders: Copyright Implications for Authors in the Digital Economy. In I. Oncioiu (Ed.), *Driving Innovation and Business Success in the Digital Economy* (pp. 83–98). Hershey, PA: IGI Global. doi:10.4018/978-1-5225-1779-5.ch007

Caravias, V. (2017). Teachers Conceptions and Approaches to Blended Learning: A Literature Review. In I. Oncioiu (Ed.), *Driving Innovation and Business Success in the Digital Economy* (pp. 228–250). Hershey, PA: IGI Global. doi:10.4018/978-1-5225-1779-5.ch015

Chen, L., Lu, M., & Tu, Y. (2017). After Auctions Complete: What Will Buyers Do Next? - A Case Study of Feedback Rating at eBay. *International Journal of E-Business Research, 13*(3), 39–56. doi:10.4018/IJEBR.2017070101

Choudhury, M. A. (2017). Cybernetic Approach for the Stock Market: An Empirical Study of Bangladesh. In I. Oncioiu (Ed.), *Driving Innovation and Business Success in the Digital Economy* (pp. 193–210). Hershey, PA: IGI Global. doi:10.4018/978-1-5225-1779-5.ch013

Constantino, G. D., & Raffaghelli, J. E. (2017). Perspectives on Digital Scholarship: Exploring Conceptual Landscape and Practices in Latin America. In A. Esposito (Ed.), *Research 2.0 and the Impact of Digital Technologies on Scholarly Inquiry* (pp. 69–96). Hershey, PA: IGI Global. doi:10.4018/978-1-5225-0830-4.ch005

Cosner, S., Tozer, S., & Zavitkovsky, P. (2016). Enacting a Cycle of Inquiry Capstone Research Project in Doctoral-Level Leadership Preparation. In V. Storey, & K. Hesbol (Eds.), *Contemporary Approaches to Dissertation Development and Research Methods* (pp. 162-183). Hershey, PA: IGI Global. doi:10.4018/978-1-5225-0445-0.ch011

Cranton, P. (2015). Teachers as Researchers: Participatory and Action Research. In V. Wang (Ed.), *Handbook of Research on Scholarly Publishing and Research Methods* (pp. 312–328). Hershey, PA: IGI Global. doi:10.4018/978-1-4666-7409-7.ch016

Crow, R., Lomotey, K., & Topolka-Jorissen, K. (2016). An Adaptive Model for a Rigorous Professional Practice Doctorate: The Disquisition. In V. Storey, & K. Hesbol (Eds.), *Contemporary Approaches to Dissertation Development and Research Methods* (pp. 205-220). Hershey, PA: IGI Global. doi:10.4018/978-1-5225-0445-0.ch013

Dawson, K., & Kumar, S. (2016). Guiding Principles for Quality Professional Practice Dissertations. In V. Storey, & K. Hesbol (Eds.), *Contemporary Approaches to Dissertation Development and Research Methods* (pp. 133-145). Hershey, PA: IGI Global. doi:10.4018/978-1-5225-0445-0.ch009

Devlin, M. (2017). An Approach to Improving Teaching in Higher Education: A Case Study Informed by the Neo-Positivist Research Paradigm. In L. Ling & P. Ling (Eds.), *Methods and Paradigms in Education Research* (pp. 68–87). Hershey, PA: IGI Global. doi:10.4018/978-1-5225-1738-2.ch005

Drago-Severson, E., Maslin-Ostrowski, P., Asghar, A., & Gaylor, S. S. (2015). Learning and Teaching Qualitative Data Analysis in a US University: Creating Supports and Scaffolds for Researcher Development. In V. Wang (Ed.), *Handbook of Research on Scholarly Publishing and Research Methods* (pp. 180–204). Hershey, PA: IGI Global. doi:10.4018/978-1-4666-7409-7.ch010

Encinas-Ferrer, C. (2017). Currency Parity and Competitiveness: The Case of Greece. In A. Vlachvei, O. Notta, K. Karantininis, & N. Tsounis (Eds.), *Factors Affecting Firm Competitiveness and Performance in the Modern Business World* (pp. 282–299). Hershey, PA: IGI Global. doi:10.4018/978-1-5225-0843-4.ch010

Esposito, A. (2017). Self-Organizing the Scholarly Practices: How the PhD Researchers Use Web 2.0 and Social Media. In A. Esposito (Ed.), *Research 2.0 and the Impact of Digital Technologies on Scholarly Inquiry* (pp. 144–166). Hershey, PA: IGI Global. doi:10.4018/978-1-5225-0830-4.ch008

Even, A., Parmet, Y., & Erez, L. (2015). Factors that Affect Customers Readiness for Internet-based BI Services. *International Journal of Business Intelligence Research*, *6*(1), 30–48. doi:10.4018/IJBIR.2015010103

Farmer, L. S. (2015). Information Architecture and the Comic Arts: Knowledge Structure and Access. In V. Wang (Ed.), *Handbook of Research on Scholarly Publishing and Research Methods* (pp. 252–271). Hershey, PA: IGI Global. doi:10.4018/978-1-4666-7409-7.ch013

Fraser, K., & Pechenkina, E. (2017). Research Paradigms Underpinning SoTL Papers: A Comparative Analysis of Two Journals. In L. Ling & P. Ling (Eds.), *Methods and Paradigms in Education Research* (pp. 42–52). Hershey, PA: IGI Global. doi:10.4018/978-1-5225-1738-2.ch003

Garwood, K. C., & Strandberg, A. G. (2016). Measuring Agreement Among Ranks: Sustainability Application. *International Journal of Business Intelligence Research*, *7*(1), 45–62. doi:10.4018/IJBIR.2016010104

Gençer, M., & Oba, B. (2017). Taming of "Openness" in Software Innovation Systems. In I. Oncioiu (Ed.), *Driving Innovation and Business Success in the Digital Economy* (pp. 26–40). Hershey, PA: IGI Global. doi:10.4018/978-1-5225-1779-5.ch003

Gendron, M., Hutwelker, C., & Kisz, K. (2016). Best Practices for Culturally Sensitive Data Visualizations. *International Journal of Business Intelligence Research*, *7*(2), 1–19. doi:10.4018/IJBIR.2016070101

Ger, T., Chu, Y., & Chang, M. (2017). Comprehension of Technology in Parent-Child Activities Using Bloom's Taxonomy of the Cognitive Domain. In I. Oncioiu (Ed.), *Driving Innovation and Business Success in the Digital Economy* (pp. 212–227). Hershey, PA: IGI Global. doi:10.4018/978-1-5225-1779-5.ch014

Giridharan, B. (2017). Education Research in an Intercultural Context: An Instance of Interpretation, Induction, and Deduction. In L. Ling & P. Ling (Eds.), *Methods and Paradigms in Education Research* (pp. 263–277). Hershey, PA: IGI Global. doi:10.4018/978-1-5225-1738-2.ch016

Gupta, V., Khanna, S., & Kim, I. (2014). Personal Financial Aggregation and Social Media Mining: A New Framework for Actionable Financial Business Intelligence (AFBI). *International Journal of Business Intelligence Research*, *5*(4), 14–25. doi:10.4018/IJBIR.2014100102

Halaweh, M., & Al Qaisi, H. (2016). Adoption of Near Field Communication (NFC) for Mobile Payments in the UAE: A Merchants Perspective. *International Journal of E-Business Research*, *12*(4), 38–56. doi:10.4018/IJEBR.2016100103

Hanafizadeh, P., & Hashemi, A. (2014). The Effect of Behavioral Factors on Stock Price Prediction using Generalized Regression and Backpropagation Neural Networks Models. *International Journal of Business Intelligence Research*, *5*(4), 44–57. doi:10.4018/IJBIR.2014100104

Hansman, C. A. (2015). Navigators on the Research Path: Teaching and Mentoring Student Qualitative Researchers. In V. Wang (Ed.), *Handbook of Research on Scholarly Publishing and Research Methods* (pp. 155–179). Hershey, PA: IGI Global. doi:10.4018/978-1-4666-7409-7.ch009

Homata, A., Mihiotis, A., & Tzortzaki, A. M. (2017). Franchise Management and the Greek Franchise Industry. In A. Vlachvei, O. Notta, K. Karantininis, & N. Tsounis (Eds.), *Factors Affecting Firm Competitiveness and Performance in the Modern Business World* (pp. 251–281). Hershey, PA: IGI Global. doi:10.4018/978-1-5225-0843-4.ch009

Israel, M. S., Sostak, S., Stewart, F. P., & Bazzi-Moughania, A. (2016). Creating an EdD Structure, program, and process fulfilling the needs of doctoral candidates and the communities they will serve. In V. Storey, & K. Hesbol (Eds.), *Contemporary Approaches to Dissertation Development and Research Methods* (pp. 70-87). Hershey, PA: IGI Global. doi:10.4018/978-1-5225-0445-0.ch006

Jamil, G. L. (2014). Why Quality? Why Value? Is it Information Related to These Aspects? In G. Jamil, A. Malheiro, & F. Ribeiro (Eds.), *Rethinking the Conceptual Base for New Practical Applications in Information Value and Quality* (pp. 1–18). Hershey, PA: IGI Global. doi:10.4018/978-1-4666-4562-2.ch001

Karantonis, I., Baourakis, G., & Zopounidis, C. (2017). Performance Evaluation of Hotel Enterprises in Crete: A Multi-Criteria Approach. In A. Vlachvei, O. Notta, K. Karantininis, & N. Tsounis (Eds.), *Factors Affecting Firm Competitiveness and Performance in the Modern Business World* (pp. 129–161). Hershey, PA: IGI Global. doi:10.4018/978-1-5225-0843-4.ch005

Kavoura, A., & Koziol, L. (2017). Polish Firms' Innovation Capability for Competitiveness via Information Technologies and Social Media Implementation. In A. Vlachvei, O. Notta, K. Karantininis, & N. Tsounis (Eds.), *Factors Affecting Firm Competitiveness and Performance in the Modern Business World* (pp. 191–222). Hershey, PA: IGI Global. doi:10.4018/978-1-5225-0843-4.ch007

Kell, C., & Czerniewicz, L. (2017). Visibility of Scholarly Research and Changing Research Communication Practices: A Case Study from Namibia. In A. Esposito (Ed.), *Research 2.0 and the Impact of Digital Technologies on Scholarly Inquiry* (pp. 97–116). Hershey, PA: IGI Global. doi:10.4018/978-1-5225-0830-4.ch006

Kelly, M. (2017). Teachers' Ontological and Epistemological Beliefs: Their Impact on Approaches to Teaching. In L. Ling & P. Ling (Eds.), *Methods and Paradigms in Education Research* (pp. 144–166). Hershey, PA: IGI Global. doi:10.4018/978-1-5225-1738-2.ch009

Khaled, R. (2017). Impact on Agricultural Sustainability of Maghreb Countries: An Empirical Analysis by 3SLS. In I. Oncioiu (Ed.), *Driving Innovation and Business Success in the Digital Economy* (pp. 66–81). Hershey, PA: IGI Global. doi:10.4018/978-1-5225-1779-5.ch006

King, S. O. (2017). Engineering Students' Approaches to Learning Mathematics. In L. Ling & P. Ling (Eds.), *Methods and Paradigms in Education Research* (pp. 167–190). Hershey, PA: IGI Global. doi:10.4018/978-1-5225-1738-2.ch010

Kochhar-Bryant, C. A. (2016). Identity, Commitment, and Change Agency: Bedrock for Bridging Theory and Practice in Doctoral Education. In V. Storey, & K. Hesbol (Eds.), *Contemporary Approaches to Dissertation Development and Research Methods* (pp. 29-42). Hershey, PA: IGI Global. doi:10.4018/978-1-5225-0445-0.ch003

Koltay, T., & Špiranec, S. (2017). Libraries Meet Research 2.0: Literacies and Services. In A. Esposito (Ed.), *Research 2.0 and the Impact of Digital Technologies on Scholarly Inquiry* (pp. 32–52). Hershey, PA: IGI Global. doi:10.4018/978-1-5225-0830-4.ch003

Laine, P., & Parkkari, P. (2017). Implications of the Strategic Agency of Sociomaterial Configurations for Participation in Strategy-Making. In I. Oncioiu (Ed.), *Driving Innovation and Business Success in the Digital Economy* (pp. 172–192). Hershey, PA: IGI Global. doi:10.4018/978-1-5225-1779-5.ch012

Lang, C. (2017). Paradigm Surfing: Cross-Disciplinary Education-Focused Research. In L. Ling & P. Ling (Eds.), *Methods and Paradigms in Education Research* (pp. 234–245). Hershey, PA: IGI Global. doi:10.4018/978-1-5225-1738-2.ch014

Lauzon, A. (2015). The Civic University, the Engaged Scholar: Implications for Scholarly Work. In V. Wang (Ed.), *Handbook of Research on Scholarly Publishing and Research Methods* (pp. 105–124). Hershey, PA: IGI Global. doi:10.4018/978-1-4666-7409-7.ch006

Lavassani, K. M., & Movahedi, B. (2017). Applications Driven Information Systems: Beyond Networks toward Business Ecosystems. In I. Oncioiu (Ed.), *Driving Innovation and Business Success in the Digital Economy* (pp. 159–171). Hershey, PA: IGI Global. doi:10.4018/978-1-5225-1779-5.ch011

Lawrence, R. L. (2015). Dancing with the Data: Arts-Based Qualitative Research. In V. Wang (Ed.), *Handbook of Research on Scholarly Publishing and Research Methods* (pp. 141–154). Hershey, PA: IGI Global. doi:10.4018/978-1-4666-7409-7.ch008

Leder, G. C. (2017). Investigating School Mathematics Performance and Affect: A Critique of Research Methods and Instruments. In L. Ling & P. Ling (Eds.), *Methods and Paradigms in Education Research* (pp. 108–123). Hershey, PA: IGI Global. doi:10.4018/978-1-5225-1738-2.ch007

Lester, S. (2016). Practice as Research: Developing the Workplace Project. In V. Storey, & K. Hesbol (Eds.), *Contemporary Approaches to Dissertation Development and Research Methods* (pp. 116-132). Hershey, PA: IGI Global. doi:10.4018/978-1-5225-0445-0.ch008

Levä, T., Ilaghi, M., Looga, V., Komu, M., Beijar, N., & Mazhelis, O. (2017). A Techno-Economic Perspective of Constrained Application Protocol. In I. Oncioiu (Ed.), *Driving Innovation and Business Success in the Digital Economy* (pp. 251–267). Hershey, PA: IGI Global. doi:10.4018/978-1-5225-1779-5.ch016

Li, J., Greenhow, C., & Askari, E. (2017). Scholars in the Digital Age: Social Scholarship and Practices. In A. Esposito (Ed.), *Research 2.0 and the Impact of Digital Technologies on Scholarly Inquiry* (pp. 1–16). Hershey, PA: IGI Global. doi:10.4018/978-1-5225-0830-4.ch001

Lima de Magalhães, J., & Quoniam, L. (2014). Perception of the Information Value for Public Health: A Case Study for Neglected Diseases. In G. Jamil, A. Malheiro, & F. Ribeiro (Eds.), *Rethinking the Conceptual Base for New Practical Applications in Information Value and Quality* (pp. 211–232). Hershey, PA: IGI Global. doi:10.4018/978-1-4666-4562-2.ch009

Lin, L., Cranton, P., & Lee, J. (2015). Research Methodologies for Multitasking Studies. In V. Wang (Ed.), *Handbook of Research on Scholarly Publishing and Research Methods* (pp. 329–348). Hershey, PA: IGI Global. doi:10.4018/978-1-4666-7409-7.ch017

Ling, L. (2017). The Power of the Paradigm: Methods and Paradigms in Education Research. In L. Ling & P. Ling (Eds.), *Methods and Paradigms in Education Research* (pp. 19–41). Hershey, PA: IGI Global. doi:10.4018/978-1-5225-1738-2.ch002

Ling, L., & Ling, P. (2017). Conclusion: Paradigm Paradiddle. In L. Ling & P. Ling (Eds.), *Methods and Paradigms in Education Research* (pp. 345–351). Hershey, PA: IGI Global. doi:10.4018/978-1-5225-1738-2.ch021

Ling, P. (2017). Transnational Education Research and Development: Paradigm Possibilities. In L. Ling & P. Ling (Eds.), *Methods and Paradigms in Education Research* (pp. 220–232). Hershey, PA: IGI Global. doi:10.4018/978-1-5225-1738-2.ch013

Ling, P., & Ling, L. (2017). Introduction: Employing Paradigms in Education Research. In L. Ling & P. Ling (Eds.), *Methods and Paradigms in Education Research* (pp. 1–18). Hershey, PA: IGI Global. doi:10.4018/978-1-5225-1738-2.ch001

Livingston, K. (2017). Undertaking Commissioned Research in Education: Do Research Paradigms Matter? In L. Ling & P. Ling (Eds.), *Methods and Paradigms in Education Research* (pp. 206–219). Hershey, PA: IGI Global. doi:10.4018/978-1-5225-1738-2.ch012

MacGregor, C. J., & Fellabaum, J. (2016). Dissertation Redesign for Scholarly Practitioners in Educational Leadership: Increasing Impact through Dissemination-Ready Sections. In V. Storey, & K. Hesbol (Eds.), *Contemporary Approaches to Dissertation Development and Research Methods* (pp. 53-69). Hershey, PA: IGI Global. doi:10.4018/978-1-5225-0445-0.ch005

Mackenzie, N. (2017). Researching the Learning and Teaching of Writing: A Retrospective Analysis of Paradigms Employed. In L. Ling & P. Ling (Eds.), *Methods and Paradigms in Education Research* (pp. 246–262). Hershey, PA: IGI Global. doi:10.4018/978-1-5225-1738-2.ch015

Malheiro da Silva, A., & Ribeiro, F. (2014). Information Evaluation: A Scientific and Comprehensive Approach. In G. Jamil, A. Malheiro, & F. Ribeiro (Eds.), *Rethinking the Conceptual Base for New Practical Applications in Information Value and Quality* (pp. 19–41). Hershey, PA: IGI Global. doi:10.4018/978-1-4666-4562-2.ch002

Manca, S., & Ranieri, M. (2017). Exploring Digital Scholarship: A Study on Use of Social Media for Scholarly Communication among Italian Academics. In A. Esposito (Ed.), *Research 2.0 and the Impact of Digital Technologies on Scholarly Inquiry* (pp. 117–142). Hershey, PA: IGI Global. doi:10.4018/978-1-5225-0830-4.ch007

Maravilhas, S. (2014). Quality Improves the Value of Patent Information to Promote Innovation. In G. Jamil, A. Malheiro, & F. Ribeiro (Eds.), *Rethinking the Conceptual Base for New Practical Applications in Information Value and Quality* (pp. 61–82). Hershey, PA: IGI Global. doi:10.4018/978-1-4666-4562-2.ch004

Marques, M. B. (2014). The Value of Information and Information Services in Knowledge Society. In G. Jamil, A. Malheiro, & F. Ribeiro (Eds.), *Rethinking the Conceptual Base for New Practical Applications in Information Value and Quality* (pp. 134–161). Hershey, PA: IGI Global. doi:10.4018/978-1-4666-4562-2.ch007

Marshall, S. M., & Klocko, B. A. (2016). Supporting Scholar-Practitioners toward Doctoral Degree Attainment: Intentional Strategies for Dissertation Completion. In V. Storey, & K. Hesbol (Eds.), *Contemporary Approaches to Dissertation Development and Research Methods* (pp. 286-299). Hershey, PA: IGI Global. doi:10.4018/978-1-5225-0445-0.ch018

McBride, N. (2015). Virtuous Business Intelligence. *International Journal of Business Intelligence Research, 6*(2), 1–17. doi:10.4018/IJBIR.2015070101

McPherson, M. J., & Lemon, N. (2017). The Hook, Woo, and Spin: Academics Creating Relations on Social Media. In A. Esposito (Ed.), *Research 2.0 and the Impact of Digital Technologies on Scholarly Inquiry* (pp. 167–187). Hershey, PA: IGI Global. doi:10.4018/978-1-5225-0830-4.ch009

Merriam, S. B. (2015). Qualitative Research: Designing, Implementing, and Publishing a Study. In V. Wang (Ed.), *Handbook of Research on Scholarly Publishing and Research Methods* (pp. 125–140). Hershey, PA: IGI Global. doi:10.4018/978-1-4666-7409-7.ch007

Nacarelli, V. (2014). Application of Six Sigma to Achieve Sales Lifecycle Optimization. *International Journal of Business Intelligence Research, 5*(4), 1–13. doi:10.4018/IJBIR.2014100101

Naghizadeh, M., Movahhed, M. K., & Jedari, Y. G. (2017). Quality Assessment Framework for Mobile Health Systems. *International Journal of E-Business Research, 13*(3), 71–85. doi:10.4018/IJEBR.2017070104

Norris, S. E. (2015). Learning and Knowledge Creation under Perpetual Construction: A Complex Responsive Approach to Applied Business Research. In V. Wang (Ed.), *Handbook of Research on Scholarly Publishing and Research Methods* (pp. 205–230). Hershey, PA: IGI Global. doi:10.4018/978-1-4666-7409-7.ch011

O'Donnell, E., & O'Donnell, L. (2015). Technology-Enhanced Learning: Towards Providing Supports for PhD Students and Researchers in Higher Education. In V. Wang (Ed.), *Handbook of Research on Scholarly Publishing and Research Methods* (pp. 231–251). Hershey, PA: IGI Global. doi:10.4018/978-1-4666-7409-7.ch012

O'Dwyer, S. T., McDonough, S. L., Jefferson, R., Goff, J. A., & Redman-MacLaren, M. (2017). Writing Groups in the Digital Age: A Case Study Analysis of Shut Up & Write Tuesdays. In A. Esposito (Ed.), *Research 2.0 and the Impact of Digital Technologies on Scholarly Inquiry* (pp. 249–269). Hershey, PA: IGI Global. doi:10.4018/978-1-5225-0830-4.ch013

Oju, O. (2017). Impact of Innovation on the Entrepreneurial Success in Selected Business Enterprises in South-West Nigeria. In I. Oncioiu (Ed.), *Driving Innovation and Business Success in the Digital Economy* (pp. 16–25). Hershey, PA: IGI Global. doi:10.4018/978-1-5225-1779-5.ch002

Oliveira, N. R., & Morgado, L. (2017). Digital Identity of Researchers 2.0: The Case of Their Personal Learning Network. In A. Esposito (Ed.), *Research 2.0 and the Impact of Digital Technologies on Scholarly Inquiry* (pp. 270–289). Hershey, PA: IGI Global. doi:10.4018/978-1-5225-0830-4.ch014

Oliveira, T., Páscoa, C., & Tribolet, J. (2014). Strategy Map: Key Instrument for Organizational Performance. In G. Jamil, A. Malheiro, & F. Ribeiro (Eds.), *Rethinking the Conceptual Base for New Practical Applications in Information Value and Quality* (pp. 250–265). Hershey, PA: IGI Global. doi:10.4018/978-1-4666-4562-2.ch011

Osimo, D., Priego, L. P., & Vuorikari, R. (2017). Alternative Research Funding Mechanisms: Make Funding Fit for Science 2.0. In A. Esposito (Ed.), *Research 2.0 and the Impact of Digital Technologies on Scholarly Inquiry* (pp. 53–67). Hershey, PA: IGI Global. doi:10.4018/978-1-5225-0830-4.ch004

Othenin-Girard, C., Guillemette, M. G., Foley, E., & Caron, C. (2015). When Competitive Intelligence Meets Geospatial Intelligence. *International Journal of Business Intelligence Research, 6*(2), 33–54. doi:10.4018/IJBIR.2015070103

Ozuem, W., & Lancaster, G. (2015). Examining the Dynamics of Value Propositions in Digital Books: A Social Constructivist Perspective. In V. Wang (Ed.), *Handbook of Research on Scholarly Publishing and Research Methods* (pp. 295–311). Hershey, PA: IGI Global. doi:10.4018/978-1-4666-7409-7.ch015

Pachidi, S., & Spruit, M. (2015). The Performance Mining Method: Extracting Performance Knowledge from Software Operation Data. *International Journal of Business Intelligence Research, 6*(1), 11–29. doi:10.4018/IJBIR.2015010102

Páscoa, C., & Tribolet, J. (2014). Actor Network Theory Applied to Organizational Change: A Case Study. In G. Jamil, A. Malheiro, & F. Ribeiro (Eds.), *Rethinking the Conceptual Base for New Practical Applications in Information Value and Quality* (pp. 233–248). Hershey, PA: IGI Global. doi:10.4018/978-1-4666-4562-2.ch010

Páscoa, C., & Tribolet, J. (2014). Change and Information Value in Military Organizations' Transformation Processes. In G. Jamil, A. Malheiro, & F. Ribeiro (Eds.), *Rethinking the Conceptual Base for New Practical Applications in Information Value and Quality* (pp. 304–323). Hershey, PA: IGI Global. doi:10.4018/978-1-4666-4562-2.ch014

Páscoa, C., & Tribolet, J. (2014). Maintaining Organizational Viability and Performance: The Organizational Configuration Map. In G. Jamil, A. Malheiro, & F. Ribeiro (Eds.), *Rethinking the Conceptual Base for New Practical Applications in Information Value and Quality* (pp. 266–283). Hershey, PA: IGI Global. doi:10.4018/978-1-4666-4562-2.ch012

Passarelli, B., & Angeluci, A. C. (2014). Interactive Generation Brazil Research: Children and Teenagers using Computers, TV, Games, and Mobile Phones. In G. Jamil, A. Malheiro, & F. Ribeiro (Eds.), *Rethinking the Conceptual Base for New Practical Applications in Information Value and Quality* (pp. 284–303). Hershey, PA: IGI Global. doi:10.4018/978-1-4666-4562-2.ch013

Paulson, E. N. (2016). Adapting and Advocating for an Online EdD Program in Changing Times and "Sacred" Cultures. In V. Storey, & K. Hesbol (Eds.), *Contemporary Approaches to Dissertation Development and Research Methods* (pp. 88-114). Hershey, PA: IGI Global. doi:10.4018/978-1-5225-0445-0.ch007

Perry, J. A. (2016). The Scholarly Practitioner as Steward of the Practice. In V. Storey, & K. Hesbol (Eds.), *Contemporary Approaches to Dissertation Development and Research Methods* (pp. 300-313). Hershey, PA: IGI Global. doi:10.4018/978-1-5225-0445-0.ch019

Pessoa, C. R., Silva, U. P., & Cruz, C. H. (2014). Information Management in Industrial Areas: A Knowledge Management View. In G. Jamil, A. Malheiro, & F. Ribeiro (Eds.), *Rethinking the Conceptual Base for New Practical Applications in Information Value and Quality* (pp. 378–395). Hershey, PA: IGI Global. doi:10.4018/978-1-4666-4562-2.ch017

Pestana, O. (2014). Information Value and Quality for the Health Sector: A Case Study of Search Strategies for Optimal Information Retrieval. In G. Jamil, A. Malheiro, & F. Ribeiro (Eds.), *Rethinking the Conceptual Base for New Practical Applications in Information Value and Quality* (pp. 116–133). Hershey, PA: IGI Global. doi:10.4018/978-1-4666-4562-2.ch006

Pogrow, S. (2016). Conducting a Design-Based Research EdD Dissertation. In V. Storey, & K. Hesbol (Eds.), *Contemporary Approaches to Dissertation Development and Research Methods* (pp. 184-204). Hershey, PA: IGI Global. doi:10.4018/978-1-5225-0445-0.ch012

Pollock, K., & Hauseman, D. C. (2017). Observational Research on the Work of School Principals: To Time or Not to Time. In L. Ling & P. Ling (Eds.), *Methods and Paradigms in Education Research* (pp. 88–107). Hershey, PA: IGI Global. doi:10.4018/978-1-5225-1738-2.ch006

Ponte, D., & Klein, S. (2017). Research and Web 2.0: Technology, Innovation, and Actor Constellations. In A. Esposito (Ed.), *Research 2.0 and the Impact of Digital Technologies on Scholarly Inquiry* (pp. 17–31). Hershey, PA: IGI Global. doi:10.4018/978-1-5225-0830-4.ch002

Prokop, J. (2017). Firm Performance and Research and Development. In A. Vlachvei, O. Notta, K. Karantininis, & N. Tsounis (Eds.), *Factors Affecting Firm Competitiveness and Performance in the Modern Business World* (pp. 108–128). Hershey, PA: IGI Global. doi:10.4018/978-1-5225-0843-4.ch004

Rahman, N., & Iverson, S. (2015). Big Data Business Intelligence in Bank Risk Analysis. *International Journal of Business Intelligence Research, 6*(2), 55–77. doi:10.4018/IJBIR.2015070104

Randolph, A. B., & Burkhalter, J. N. (2016). From General to Specific: Understanding Individual Characteristics and their Relationship with Neural Recordings during Media Consumption. *International Journal of Business Intelligence Research, 7*(2), 32–49. doi:10.4018/IJBIR.2016070103

Rascão, J. P. (2014). Information Management in Complex and Turbulent Environments. In G. Jamil, A. Malheiro, & F. Ribeiro (Eds.), *Rethinking the Conceptual Base for New Practical Applications in Information Value and Quality* (pp. 349–377). Hershey, PA: IGI Global. doi:10.4018/978-1-4666-4562-2.ch016

Ravitch, S. M., & Lytle, S. L. (2016). Becoming Practitioner-Scholars: The Role of Practice-Based Inquiry Dissertations in the Development of Educational Leaders. In V. Storey, & K. Hesbol (Eds.), *Contemporary Approaches to Dissertation Development and Research Methods* (pp. 146-161). Hershey, PA: IGI Global. doi:10.4018/978-1-5225-0445-0.ch010

Richards, G. S. (2016). Business Intelligence and Analytics Research: A Peek Inside the Black Box. *International Journal of Business Intelligence Research, 7*(1), 1–10. doi:10.4018/IJBIR.2016010101

Robertson, M. E. (2017). Doctoral Research Supervision: Interpretive, Developmental, Transformative, and Culturally Adaptive. In L. Ling & P. Ling (Eds.), *Methods and Paradigms in Education Research* (pp. 53–66). Hershey, PA: IGI Global. doi:10.4018/978-1-5225-1738-2.ch004

Rocco, T. S., Gionti, L. A., Januszka, C. M., Munn, S. L., & Collins, J. C. (2015). Creating a Support Structure for Academic Writing and Publication Support: The Rationale and Lessons Learned. In V. Wang (Ed.), *Handbook of Research on Scholarly Publishing and Research Methods* (pp. 21–37). Hershey, PA: IGI Global. doi:10.4018/978-1-4666-7409-7.ch002

Rockembach, M. (2014). New Approaches to Information Quality in the Digital Environment: Clues, Evidence, and Proof. In G. Jamil, A. Malheiro, & F. Ribeiro (Eds.), *Rethinking the Conceptual Base for New Practical Applications in Information Value and Quality* (pp. 43–60). Hershey, PA: IGI Global. doi:10.4018/978-1-4666-4562-2.ch003

Rose, M. (2017). The Black Academy: A Renaissance Seen Through a Paradigmatic Prism. In L. Ling & P. Ling (Eds.), *Methods and Paradigms in Education Research* (pp. 326–343). Hershey, PA: IGI Global. doi:10.4018/978-1-5225-1738-2.ch020

Rosenzweig, E. D., & Bendoly, E. (2017). An Investigation of Competitor Networks in Manufacturing Strategy and Implications for Performance. In A. Vlachvei, O. Notta, K. Karantininis, & N. Tsounis (Eds.), *Factors Affecting Firm Competitiveness and Performance in the Modern Business World* (pp. 43–82). Hershey, PA: IGI Global. doi:10.4018/978-1-5225-0843-4.ch002

Rouhani, S., & Savoji, S. R. (2016). A Success Assessment Model for BI Tools Implementation: An Empirical Study of Banking Industry. *International Journal of Business Intelligence Research*, 7(1), 25–44. doi:10.4018/IJBIR.2016010103

Rusko, R. (2017). Strategic Turning Points in ICT Business: The Business Development, Transformation, and Evolution in the Case of Nokia. In I. Oncioiu (Ed.), *Driving Innovation and Business Success in the Digital Economy* (pp. 1–15). Hershey, PA: IGI Global. doi:10.4018/978-1-5225-1779-5.ch001

Rusko, R., & Merenheimo, P. (2017). Co-Creating the Christmas Story: Digitalizing as a Shared Resource for a Shared Brand. In I. Oncioiu (Ed.), *Driving Innovation and Business Success in the Digital Economy* (pp. 137–157). Hershey, PA: IGI Global. doi:10.4018/978-1-5225-1779-5.ch010

Sabina, L. L., Curry, K. A., Harris, E. L., Krumm, B. L., & Vencill, V. (2016). Assessing the Performance of a Cohort-Based Model Using Domestic and International Practices. In V. Storey, & K. Hesbol (Eds.), *Contemporary Approaches to Dissertation Development and Research Methods* (pp. 256-272). Hershey, PA: IGI Global. doi:10.4018/978-1-5225-0445-0.ch016

Sedivy-Benton, A. L., & O'Kelly, M. K. (2015). Connecting Theory to Practice: Making Research Real for Graduate Students. In V. Wang (Ed.), *Handbook of Research on Scholarly Publishing and Research Methods* (pp. 38–60). Hershey, PA: IGI Global. doi:10.4018/978-1-4666-7409-7.ch003

Selkrig, M., & Keamy, (. K. (2017). Becoming Comfortable with Supercomplexity: Looking Back, Forward, In, Out, and Shaking it About!. In L. Ling, & P. Ling (Eds.), *Methods and Paradigms in Education Research* (pp. 309-325). Hershey, PA: IGI Global. doi:10.4018/978-1-5225-1738-2.ch019

Shelton, K., & Creghan, K. A. (2015). Demystifying the Delphi Method. In V. Wang (Ed.), *Handbook of Research on Scholarly Publishing and Research Methods* (pp. 375–395). Hershey, PA: IGI Global. doi:10.4018/978-1-4666-7409-7.ch019

Smith, C. (2017). Scientific Realism and the Study of Higher Education Curriculum and the Student Experience. In L. Ling & P. Ling (Eds.), *Methods and Paradigms in Education Research* (pp. 124–142). Hershey, PA: IGI Global. doi:10.4018/978-1-5225-1738-2.ch008

Spilker, M. J., Silva, M. P., & Morgado, L. (2017). Research 2.0: The Contribution of Content Curation and Academic Conferences. In A. Esposito (Ed.), *Research 2.0 and the Impact of Digital Technologies on Scholarly Inquiry* (pp. 231–248). Hershey, PA: IGI Global. doi:10.4018/978-1-5225-0830-4.ch012

Spruit, M., & de Boer, T. (2014). Business Intelligence as a Service: A Vendors Approach. *International Journal of Business Intelligence Research, 5*(4), 26–43. doi:10.4018/IJBIR.2014100103

Starr-Glass, D. (2015). Scholarship of Teaching and Learning: Promoting Publication or Encouraging Engagement? In V. Wang (Ed.), *Handbook of Research on Scholarly Publishing and Research Methods* (pp. 61–83). Hershey, PA: IGI Global. doi:10.4018/978-1-4666-7409-7.ch004

Stewart, T. J. (2016). Educational Leaders and the Program Evaluation Dissertation with Logic Model. In V. Storey, & K. Hesbol (Eds.), *Contemporary Approaches to Dissertation Development and Research Methods* (pp. 221-237). Hershey, PA: IGI Global. doi:10.4018/978-1-5225-0445-0.ch014

Storey, V. A., & Fulton, J. (2016). The Consequential Relationship between Doctoral Course Design and Capstone Design. In V. Storey, & K. Hesbol (Eds.), *Contemporary Approaches to Dissertation Development and Research Methods* (pp. 1-16). Hershey, PA: IGI Global. doi:10.4018/978-1-5225-0445-0.ch001

Swanson, K. W., West, J., Carr, S., & Augustine, S. (2015). Supporting Dissertation Writing Using a Cognitive Apprenticeship Model. In V. Wang (Ed.), *Handbook of Research on Scholarly Publishing and Research Methods* (pp. 84-104). Hershey, PA: IGI Global. doi:10.4018/978-1-4666-7409-7.ch005

Swapana, M., & Padmavathy, C. (2017). Factors Influencing Dependency on Smartphone and the Impact on Purchase Behaviour: An Empirical Research. *International Journal of E-Business Research, 13*(2), 79–92. doi:10.4018/IJEBR.2017040105

Turner, J., Petrunin, K. F., & Allen, J. (2015). Developing Multilevel Models for Research. In V. Wang (Ed.), *Handbook of Research on Scholarly Publishing and Research Methods* (pp. 463–489). Hershey, PA: IGI Global. doi:10.4018/978-1-4666-7409-7.ch023

Vargas-Hernández, J. G., Ioannis, A. I., & González-Armenta, L. (2017). Joint Venture Efficiency through Skills Complementarity or by Reducing Transaction Costs?: A Case Study of an Apparel Company in an Emerging Market. In A. Vlachvei, O. Notta, K. Karantininis, & N. Tsounis (Eds.), *Factors Affecting Firm Competitiveness and Performance in the Modern Business World* (pp. 162–190). Hershey, PA: IGI Global. doi:10.4018/978-1-5225-0843-4.ch006

Verbitskiy, Y., & Yeoh, W. (2016). An End-User Metadata Model on Object and Element Levels for Business Intelligence Users. *International Journal of Business Intelligence Research, 7*(2), 50–58. doi:10.4018/IJBIR.2016070104

Vicars, M. (2017). A Code of Our Own: Making Meaning Queerly. In L. Ling & P. Ling (Eds.), *Methods and Paradigms in Education Research* (pp. 191–204). Hershey, PA: IGI Global. doi:10.4018/978-1-5225-1738-2.ch011

Vlachvei, A., & Notta, O. (2017). Firm Competitiveness: Theories, Evidence, and Measurement. In A. Vlachvei, O. Notta, K. Karantininis, & N. Tsounis (Eds.), *Factors Affecting Firm Competitiveness and Performance in the Modern Business World* (pp. 1–42). Hershey, PA: IGI Global. doi:10.4018/978-1-5225-0843-4.ch001

Wahi, A. K., & Misra, R. K. (2016). Success in the Web 2.0 Ecosystem: Exploring the State and Determinants in Indian Businesses. *International Journal of E-Business Research, 12*(1), 19–42. doi:10.4018/IJEBR.2016010102

Wang, T., Verbitskiy, Y., & Yeoh, W. (2016). Depicting Data Quality Issues in Business Intelligence Environment through a Metadata Framework. *International Journal of Business Intelligence Research, 7*(2), 20–31. doi:10.4018/IJBIR.2016070102

Watson, D. L., & Nehls, K. (2016). Alternative Dissertation Formats: Preparing Scholars for the Academy and Beyond. In V. Storey, & K. Hesbol (Eds.), *Contemporary Approaches to Dissertation Development and Research Methods* (pp. 43-52). Hershey, PA: IGI Global. doi:10.4018/978-1-5225-0445-0.ch004

Wen-Jang (Kenny) Jih, & Lee, S. (2017). Effects of Social Presence and Social Capital on User Loyalty to the Social Networking Website: The Case of Facebook Usage in Taiwan. *International Journal of E-Business Research, 13*(3), 57–70. doi:10.4018/IJEBR.2017070104

Wynn, M. G., & Brinkmann, D. (2016). Exploiting Business Intelligence for Strategic Knowledge Management: A German Healthcare Insurance Industry Case Study. *International Journal of Business Intelligence Research, 7*(1), 11–24. doi:10.4018/IJBIR.2016010102

Xu, L. (2015). Measurement Development and Validation in Research: Statistical Techniques and Illustrations. In V. Wang (Ed.), *Handbook of Research on Scholarly Publishing and Research Methods* (pp. 435–462). Hershey, PA: IGI Global. doi:10.4018/978-1-4666-7409-7.ch022

Yadav, V. (2017). Sustainable Development Challenges in Developing Countries: Can Technology Provide Inclusive Solutions? In I. Oncioiu (Ed.), *Driving Innovation and Business Success in the Digital Economy* (pp. 56–65). Hershey, PA: IGI Global. doi:10.4018/978-1-5225-1779-5.ch005

Yap, C. S., Ong, M. Y., & Ahmad, R. (2017). Online Product Review, Product Knowledge, Attitude, and Online Purchase Behavior. *International Journal of E-Business Research, 13*(3), 1–19. doi:10.4018/IJEBR.2017070103

Yu, K., Mustapha, N., & Oozeer, N. (2017). Google Scholar's Filter Bubble: An Inflated Actuality? In A. Esposito (Ed.), *Research 2.0 and the Impact of Digital Technologies on Scholarly Inquiry* (pp. 211–229). Hershey, PA: IGI Global. doi:10.4018/978-1-5225-0830-4.ch011

Zambo, D., & Zambo, R. (2016). The Role of Theory in EdD Programs and Dissertations in Practice. In V. Storey, & K. Hesbol (Eds.), *Contemporary Approaches to Dissertation Development and Research Methods* (pp. 17-28). Hershey, PA: IGI Global. doi:10.4018/978-1-5225-0445-0.ch002

Zaninelli, T. (2014). The Communication and Information Role on the Innovation Process Performance in the Collaborative Context: New Service Project Cases. In G. Jamil, A. Malheiro, & F. Ribeiro (Eds.), *Rethinking the Conceptual Base for New Practical Applications in Information Value and Quality* (pp. 162–210). Hershey, PA: IGI Global. doi:10.4018/978-1-4666-4562-2.ch008

Zhang, S. (2016). Principles of Discussion in Qualitative Educational Research. In V. Storey, & K. Hesbol (Eds.), *Contemporary Approaches to Dissertation Development and Research Methods* (pp. 273-285). Hershey, PA: IGI Global. doi:10.4018/978-1-5225-0445-0.ch017

Zhu, Y. (2017). Academics' Active and Passive Use of YouTube for Research and Leisure. In A. Esposito (Ed.), *Research 2.0 and the Impact of Digital Technologies on Scholarly Inquiry* (pp. 188–210). Hershey, PA: IGI Global. doi:10.4018/978-1-5225-0830-4.ch010

Zolfaghari, S., Keyvanpour, M. R., & Zall, R. (2017). Analytical Review on Ontological Human Activity Recognition Approaches. *International Journal of E-Business Research, 13*(2), 58–78. doi:10.4018/IJEBR.2017040104

About the Author

Amiram Porath has been involved in Innovation projects and activities for over 25 years. He managed two start-ups in technology incubators, other start-ups under private financing, established the EU office at Tel Aviv University where he developed national and international R&D consortia. His PhD in management focused on the R&D cooperation between Industry and Academy. He has led standardization efforts regarding critical resources management as part of his environmental activities and has been consulting CREST and the Israeli Chief Scientist Office regarding the cooperation between industry and Academy. Familiarity with the outlooks of Industry, Academy and Government as well as national and international innovation systems is pushing forward his writing. For the last three years he has been focused on the management of three national programs, as well as the environmental activity and entrepreneurship coaching. In 2008, he received the title "Cavaliere" from the Italian Republic.

Index

www.ingramcontent.com/pod-product-compliance
Lightning Source LLC
LaVergne TN
LVHW080927200926
848479LV00005B/106